Lecture Notes in Computer Scie

T0237877

Commenced Publication in 1973
Founding and Former Series Editors:
Gerhard Goos, Juris Hartmanis, and Jan van Leeuwen

Zeno J. M. H. Geradts   Katrin Y. Franke
Cor J. Veenman (Eds.)

Computational Forensics

Third International Workshop, IWCF 2009
The Hague, The Netherlands, August 13-14, 2009
Proceedings

 Springer

Volume Editors

Zeno J. M. H. Geradts
Netherlands Forensic Institute
The Hague, The Netherlnds
E-mail: z.geradts@nfi.monjus.nl

Katrin Y. Franke
Norwegian Information Security Laboratory
Gjøvik , Norway
E-mail: kyfranke@ieee.org

Cor J. Veenman
Netherlands Forensic Institute
The Hague, The Netherlands
E-mail: C.J.Veenman@uva.nl

Library of Congress Control Number: Applied for

CR Subject Classification (1998): I.5, I.2.10, I.4, I.7.5, I.2.7, I.6

LNCS Sublibrary: SL 6 – Image Processing, Computer Vision, Pattern Recognition, and Graphics

ISSN 0302-9743

ISBN 978-3-642-03520-3 Springer Berlin Heidelberg New York

Typesetting: Camera-ready by author, data conversion by Scientific Publishing Services, Chennai, India
Printed on acid-free paper   SPIN: 12729067   06/3180   5 4 3 2 1 0

Preface

This *Lecture Notes in Computer Science* (LNCS) volume contains the papers presented at the Third International Workshop on Computational Forensics (IWCF 2009), held August 13–14, 2009 at The Netherlands Forensic Institute in The Hague, The Netherlands.

Computational forensics is a research domain focusing on the investigation of forensic problems using computational methods. Its primary goal is the discovery and advancement of forensic knowledge involving modeling, computer simulation, and computer-based analysis and recognition in studying and solving forensic problems.

The Computational Forensics Workshop series is intended as a forum for researchers and practicioners in all areas of computational and forensic sciences. This forum discusses current challenges in computer-assisted forensics and presents recent progress and advances.

IWCF addresses a broad spectrum of forensic disciplines from criminal investigation. The NAS report "Strengthening Forensic Science in the United States: A Path Forward," published in 2009, already concluded that: "With the exception of nuclear DNA analysis, no forensic method has been rigorously shown to have the capacity to consistently and with a high degree of certainty demonstrate a connection between evidence and specific individual or source." Computational methods can assist with statistical methods using, for example, likelihood ratios to strengthen the different fields of forensic science.

The organization of such an event is not possible without the effort and the enthousiasm of everyone involved. We thank all the members of the Program Committee for the timely reviewing process and the adherence to high standards.

June 2009

Zeno J.M.H. Geradts
Katrin Y. Franke
Cor J. Veenman

IWCF 2009 Organization

IWCF 2009 was jointly organized by The Netherlands Forensic Institute, The Hague, The Netherlands and the University of Amsterdam, Amsterdam, The Netherlands.

Workshop Co-chairs

Zeno J.M.H. Geradts	Netherlands Forensic Institute, The Netherlands
Katrin Y. Franke	Gjøvik University College, Norway
Cor J. Veenman	Netherlands Forensic Institute/University of Amsterdam, The Netherlands

Program Committee

Yoshinori Akao	National Research Institute of Police Science, Japan
Faouzi Alaya Cheikh	Gjøvik University College, Norway
Lashon B. Booker	The MITRE Corporation, USA
Thomas Breuel	University of Kaiserslautern, Germany
Joseph P. Campbell	MIT Lincoln Laboratory, USA
Oscar Cordón	European Centre for Soft Computing, Spain
Edward J. Delp	Purdue University, USA
Patrick De Smet	FOD Justitie, Belgium
Andrzej Drygajlo	Swiss Federal Institute of Technology Lausanne, Switzerland
Robert P.W. Duin	Delft University of Technology, The Netherlands
Cinthia Freitas	Pontifical Catholic University of Parana, Brazil
Simson L. Garfinkel	School of Engineering and Applied Sciences, USA
Peter Gill	Strathclyde University, UK
Lawrence Hornak	West Virginia University, USA
Anil K. Jain	Michigan State University, USA
Mario Köppen	Kyushu Institute of Technology, Japan
Deborah Leben	US Secret Service, USA
Didier Meuwly	Netherlands Forensic Institute, The Netherlands
Milan Milosavljević	University of Belgrade, Serbia
Slobodan Petrović	Gjøvik University College, Norway
Olivier Ribaux	University of Lausanne, Switzerland
Hiroshi Sako	Hitachi Central Research Laboratory, Japan
Reva Schwartz	US Secret Service, USA

Table of Contents

Speech and Linguistics

Printers

Fingerprints

Visualisation

Multimedia

Handwriting

Documents

Statistical Evaluation of Biometric Evidence in Forensic Automatic Speaker Recognition

Andrzej Drygajlo

Speech Processing and Biometrics Group
Swiss Federal Institute of Technology Lausanne (EPFL)
CH-1015 Lausanne, Switzerland
{andrzej.drygajlo}@epfl.ch

Abstract. Forensic speaker recognition is the process of determining if a specific individual (suspected speaker) is the source of a questioned voice recording (trace). This paper aims at presenting forensic automatic speaker recognition (FASR) methods that provide a coherent way of quantifying and presenting recorded voice as biometric evidence. In such methods, the biometric evidence consists of the quantified degree of similarity between speaker-dependent features extracted from the trace and speaker-dependent features extracted from recorded speech of a suspect. The interpretation of recorded voice as evidence in the forensic context presents particular challenges, including within-speaker (within-source) variability and between-speakers (between-sources) variability. Consequently, FASR methods must provide a statistical evaluation which gives the court an indication of the strength of the evidence given the estimated within-source and between-sources variabilities. This paper reports on the first ENFSI evaluation campaign through a fake case, organized by the Netherlands Forensic Institute (NFI), as an example, where an automatic method using the Gaussian mixture models (GMMs) and the Bayesian interpretation (BI) framework were implemented for the forensic speaker recognition task.

1 Introduction

Biometric methods have evolved significantly and have become one of the most convincing ways to confirm the identity of an individual [1]. Several types of biometric characteristics can be used depending on their intrinsic reliability and user acceptance relating to a particular application. They include fingerprint patterns, face and iris characteristics, hand geometry, palmprints and hand vein patterns, signature dynamics and voice patterns. It is important to know that the voice patterns are not an exceptional biometric measurement as an identity trait. Despite the variety of characteristics, the systems that measure biometric differences between people have essentially the same architecture and many factors are common across several biometric processes. This generic architecture of processing and statistical modelling starts from signal sensing, passes through features extraction and their statistical modelling and ends at the stage of features against model comparison and interpretation of the comparison scores.

Z.J.M.H. Geradts, K.Y. Franke, and C.J. Veenman (Eds.): IWCF 2009, LNCS 5718, pp. 1–12, 2009.

On the other hand, forensic sciences are defined as the body of scientific principles and technical methods applied to criminal investigations to demonstrate the existence of a crime and help justice identify the authors and his modus operandi. Speaker recognition is the general term used to include all of the many different tasks of discriminating people based on the sound of their voices. Forensic speaker recognition involves the comparison of recordings of an unknown voice (questioned recording) with one or more recordings of a known voice (voice of the suspected speaker). The approaches commonly used for technical forensic speaker recognition include the aural-perceptual, auditory-instrumental, and automatic methods [2]. Forensic automatic speaker recognition (FASR) is an established term used when automatic speaker recognition methods are adapted to forensic applications [3]. Normally, in automatic speaker recognition, the statistical models of acoustic features of the speaker's voice and the acoustic features of questioned recordings are compared. FASR offers data-driven methodology for quantitative interpretation of recorded speech as evidence. It is a relatively recent application of digital speech signal processing and pattern recognition for judicial purposes and particularly law enforcement.

The goal of this paper is to present a methodological approach for forensic automatic speaker recognition (FASR) based on scientifically sound principles and to propose a practical solution based on state-of-the-art technology [4]. The interpretation of recorded speech as biometric evidence in the forensic context presents particular challenges [5]. The means proposed for dealing with this is through Bayesian inference and data-driven approach. A probabilistic model - the odds form of Bayes' theorem and likelihood ratio - seems to be an adequate tool for assisting forensic experts in the speaker recognition domain to interpret this evidence [6]. In forensic speaker recognition, statistical modelling techniques are based on the distribution of various features pertaining to the suspect's speech and its comparison to the distribution of the same features in a reference population with respect to the questioned recording.

In this paper, the state-of-the-art automatic, text-independent speaker recognition system, based on Gaussian mixture model (GMM) is used as an example. It is adapted to the Bayesian interpretation (BI) framework to estimate the within-source variability of the suspected speaker and the between-sources variability, given the questioned recording. This double-statistical approach (BI-GMM) gives an interesting solution for the interpretation of the recorded speech as evidence in the judicial process as it has been shown in the first ENFSI (European Network of Forensic Science Institute) evaluation campaign through a fake case, organized by the Netherlands Forensic Institute (NFI) [7].

The paper is structured as follows. Section 2 highlights the meaning of biometric evidence in the forensic speech recognition. In Section 3 a general Bayesian framework for interpretation of the biometric evidence of voice is introduced. Section 4 presents the ENFSI-NFI speaker recognition evaluation through a fake case using the Bayesian interpretation method. Section 5 concludes the paper.

2   Voice as Biometric Evidence

When using forensic automatic speaker recognition (FASR) the goal is to determine whether an unknown voice of a questioned recording (trace) belongs to a suspected speaker (source). The biometric evidence of voice consists of the quantified degree of similarity between speaker dependent features extracted from the trace, and speaker dependent features extracted from recorded speech of a suspect, represented by his or her model, so the evidence does not consist of the speech itself. To compute the evidence, the processing chain illustrated in Fig. 1 may be employed. However, the calculated value of evidence does not allow the forensic expert alone to make an inference on the identity of the speaker.

The most persistent real-world challenge in this field is the variability of speech. There is within-speaker (within-source) variability as well as between-speakers (between-sources) variability. Consequently, forensic speaker recognition methods should provide a statistical-probabilistic evaluation, which attempts to give the court an indication of the strength of the evidence, given the estimated within-source variability and the between-sources variability.

3   Strength of Voice Biometric Evidence

The strength of voice evidence is the result of the interpretation of the evidence, expressed in terms of the likelihood ratio of two alternative hypotheses $LR = p(E|H_0)/p(E|H_1)$.

The odds form of Bayes' theorem shows how new data (questioned recording) can be combined with prior background knowledge (prior odds (province of the court)) to give posterior odds (province of the court) for judicial outcomes or issues (Eq. 1). It allows for revision based on new information of a measure of uncertainty (likelihood ratio of the evidence (province of the forensic expert)) which is applied to the pair of competing hypotheses: H_0 - the suspected speaker is the source of the questioned recording, H_1 - the speaker at the origin of the questioned recording is not the suspected speaker:

$$\frac{p(H_0|E)}{p(H_1|E)} = \frac{p(E|H_0)}{p(E|H_1)} \cdot \frac{p(H_0)}{p(H_1)}. \tag{1}$$

The principal structure for the calculation and the interpretation of the evidence is presented in Fig. 1. It includes the collection (or selection) of the databases, the automatic speaker recognition and the Bayesian interpretation [6]. The methodological approach based on a Bayesian interpretation (BI) framework is independent of the automatic speaker recognition method chosen, but the practical solution presented in this paper as an example uses text-independent speaker recognition system based on Gaussian mixture model (GMM). The Bayesian interpretation (BI) methodology needs a two-stage statistical approach.

The first stage consists in modeling multivariate speech feature data using GMMs. The speech signal can be represented by a sequence of short-term feature vectors. This is known as feature extraction (Fig.1). It is typical to use

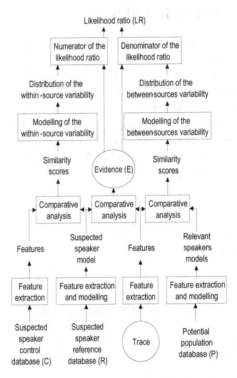

Fig. 1. Schema for the evaluation of the likelihood ratio

features based on the various speech production and perception models. Although there are no exclusive features conveying speaker identity in the speech signal, from the source-filter theory of speech production it is known that the speech spectrum envelope encodes information about the speaker's vocal tract shape. Thus some form of spectral envelope based features, e.g. RASTA-PLP, is used in most speaker recognition systems even if they are dependent on external recording conditions.

The second stage transforms the data to a univariate projection based on modeling the similarity scores. The exclusively multivariate approach is also possible but it is more difficult to articulate in layman terms such that the judge and the attorneys may understand them. The GMM method is not only used to calculate the evidence by comparing the questioned recording (trace) to the GMM of the suspected speaker (source), but it is also used to produce data necessary to model the within-source variability of the suspected speaker and the between-sources variability of the potential population of relevant speakers, given the questioned recording. The interpretation of the evidence consists of calculating the likelihood ratio using the probability density functions (pdfs) of the variabilities and the numerical value of evidence.

The information provided by the analysis of the questioned recording (trace) leads to specify the initial reference population of relevant speakers (potential population) having voices similar to the trace, and, combined with the police investigation, to focus on and select a suspected speaker. The methodology presented needs three databases for the calculation and the interpretation of the evidence: the potential population database (P), the suspected speaker reference database (R) and the suspected speaker control database (C). The purpose of the P, R and C databases is presented in Section 4.4.

The methodology presented in this section was used in the first ENFSI evaluation campaign through a fake case, organized by the Netherlands Forensic Institute (NFI).

4 ENFSI-NFI Speaker Recognition Evaluation through a Fake Case

When the Expert Working Group for Forensic Speech and Audio Analysis within the European Network of Forensic Science Institutes (ENFSI) was formed in 1997, one of its main goals was to gain insight into the different methods that are employed in the field of speaker identification within these institutes. In 2004, a collaborative evaluation exercise was constructed at the Netherlands Forensic Institute (NFI) with English material that was recorded especially for this purpose. During several meetings and conferences, in 2003 and 2004, twenty potential participants were found willing to take part in the evaluation. Twelve reports were returned by the start of 2005, together with the results of all measurements that had been done and a completed questionnaire asking about the experience of the expert, the time spent on the collaborative exercise, the software that was used, etc. In this paper, the collaborative evaluation exercise is described, and a summary of the results using automatic speaker recognition method is presented [8].

4.1 Forensic Speaker Recognition Task

Twelve audio recordings were provided, by the Netherlands Forensic Institute (NFI) as part of a fake case evaluation, consisting of two reference recordings $R1$ and $R2$, and ten questioned recordings $Q1 - Q10$. The ten questioned recordings consisted of conversations between two speakers, i.e., each containing the speech of a known speaker and an unknown speaker. The two reference recordings consisted of two conversations between a known speaker and a suspected speaker.

The aim of the analysis was to determine whether the speech of the unknown speaker in each of the questioned recordings was produced by the suspected speaker in the reference recordings.

4.2 The Audio Materials

1 CD-ROM with 12 recordings in 16 kHz, 16-bit Linear PCM wave files were provided. According to the accompanying documentation, these recordings were

recorded directly from a phone line onto a Digital Audio Tape (DAT), at 44 kHz and then down-sampled to 16 kHz and later transferred to a computer using Cool Edit Pro 2.0.2. Detailed transcriptions of the recordings with the corresponding dates, time and telephone numbers were also provided.

4.3   Investigation Work

The evaluation process constituted of the two following steps:

- **Pre-processing**: Pre-processing consisting of segmentation of the audio into the speech of individual speakers, removal of non-speech regions and selection of reference (R) and control (C) recordings was performed, in order to prepare the recordings for the analysis. It was ascertained by NFI, that all of the recordings provided were performed with a fixed telephone and that there was no mobile (GSM) channel effect in the recording conditions of all the recordings in the case. Because of this, no attempt at compensating for mismatched conditions was made.
- **Analysis**: The Bayesian interpretation methodology for the forensic automatic speaker recognition was used for this case [6]. The result of the comparison between the model of the suspect speaker and the questioned recording (called biometric evidence (E)) is evaluated given two hypotheses: H_0 - the suspect is the source of the questioned recording H_1 - anyone else in the relevant population is the source. The result of this evaluation is the strength of evidence, which is expressed as a likelihood ratio (LR).

Pre-Processing

1. **Acquisition and Down-sampling**: Acquisition was unnecessary as the files were already in digital format. However, in order to maintain consistency with the other databases used for comparison, it was necessary to down-sample the audio files to 8 kHz, 16-bit Linear PCM files using Cool Edit Pro 2.0.
2. **Segmentation**: The questioned recordings and the reference recordings were in the form of conversations between two speakers. In order to compare the speech of individual speakers it was necessary to segment each of the conversations. This segmentation was performed aurally, with the help of the transcripts provided. Zones of overlap, laughs, and other anomalies were discarded.
3. **Removal of Non-Speech Regions**: The recordings were passed through a voice activity detector (VAD), which separates speech and non-speech regions, using instantaneous signal to noise ratio (SNR). The non-speech regions of the recording contain more information about the conditions of ambient noise present in the recording and no speaker-dependent information. Removal of these non-speech regions better allows for speaker specific characteristics to be considered, when modelling voice.

4. **Selection of Reference and Control Recordings**: The two recordings $R1$ and $R2$, called by NFI reference recordings of the suspected speaker were divided into reference recordings of database R that would be used for training statistical models of his speech and control recordings of database (C) that were compared with the models of the suspected speaker, in order to estimate the within-source variability of his speech (Section 4.4).

4.4   Databases for Bayesian Interpretation

In the first part of the analysis, the Bayesian interpretation methodology presented in [6], was used as a means of calculating the strength of evidence.

This Bayesian methodology requires, in addition to the questioned recording, the use of three databases: a suspected speaker reference database (R), a suspected speaker control database (C), a potential population database (P), and the questioned recording database (T).

- The P database contains an exhaustive coverage of recordings of all possible voices satisfying the hypothesis (H_1): anyone chosen at random from a relevant population could be the source of the trace. These recordings are used to create models of speakers to evaluate the between-sources variability given the trace.
- The R database contains recordings of the suspected speaker that are as close as possible (in recording conditions and linguistically) to the recordings of speakers of P, and it is used to create the suspected speaker's model, exactly as is done with models of P.
- The C database consists of recordings of the suspected speaker that are very similar to the trace, and it is used to estimate the within-source variability of the voice of the suspected speaker.

A brief summary of the steps required in order to calculate the evidence and its strength for a given trace is as follows:

1. The trace is compared with the statistical model of the suspect (created using database R), and the resulting score is the evidence score (E).
2. The trace is compared with statistical models of all the speakers in the potential population (P). The distribution of obtained scores is an estimation of the between-sources variability of the trace with the potential population.
3. The suspected speaker control database (C) recordings are compared with the models created with the suspected speaker reference database (R) for the suspect, and the distribution of scores obtained estimates the suspect's within-source variability.
4. The likelihood ratio (LR) (i.e., the ratio of support that the evidence (E) lends to each of the hypotheses), is given by the ratio of the heights of the within-source and between-sources distributions at the point E.

In this analysis, the potential population database (P) used is the PolyCOST 250 database. We have used 73 speakers from this database in the analysis. The suspected speaker reference database consisted of two reference files of duration 2m 19s, and 2m 17s. The suspected speaker control databases consisted of 7 recordings of 20 seconds duration each. This database contains mainly European speakers who speak both in English and in their mother-tongue. This database was chosen among the available databases because it was found to be best suited to the case, especially in the language (English spoken by European speakers) and technical conditions (fixed European telephone network) under which the reference recordings of the suspect were made.

Note that an accurate estimation of the likelihood ratio in a Bayesian framework is possible, only if the technical conditions of the suspected speaker reference (R) and potential population (P) databases are identical, and the suspected speaker control database (C) was recorded in the same conditions as the questioned recording. More explicitly, following assumptions have to be satisfied: 1) the suspected speaker control database and the questioned recording were recorded in similar conditions, 2) the suspected speaker reference database and the potential population database were recorded in similar recording conditions.

In practice, it can be observed that it is very difficult to satisfy all these requirements. Incompatibilities in the databases used can result in under-estimation or over-estimation of the likelihood ratio.

4.5 Analysis

During the preprocessing step the recordings were segmented into recordings of individual speakers. Each of the segmented recordings contains only the speech of a specific speaker. This segmentation was performed aurally with the help of the transcripts provided. The set of recordings obtained, along with their durations, is presented in Table 1 .

The files R01_Peter.wav and R02_Peter.wav were further segmented into:

- two reference files R01_Peter_Ref1.wav (2m 17s) and R02_Peter_Ref1.wav (2m 19s) (R database)
- seven control recordings R01_Peter_C01.wav, R01_Peter_C02.wav, R01_Peter_C03.wav, R01_Peter_C04.wav, R02_Peter_C01.wav, R02_Peter_C02.wav and R02_Peter_C03.wav each of 20s each (C database).

The P database used is the PolyCOST 250 database . We have used 73 speakers from this database in the analysis.

The following analysis procedure was then applied to the R, C and P databases thus created:

- Analysis and creation of models of the speakers voice: Extraction of 12 RASTA-PLP features for each analysis frame and creation of a statistical model by means of a 64 component Gaussian mixture model (GMM).
- Within-source variability estimation: Comparison between statistical model of the features of the reference recording and the features of the control recordings of the suspected speaker.

Table 1. Individual speakers segments and their durations

No.	Source Original Recording	Speaker Segmented Recordings Analyzed	Length of segmented recording (s)
1	Q1.wav	Q01_Eric.wav	169.46
2	Q1.wav	Q01_NN_Male.wav	172.28
3	Q2.wav	Q02_Eric.wav	20.73
4	Q2.wav	Q02_NN_Male.wav	11.51
5	Q3.wav	Q03_Eric.wav	91.38
6	Q3.wav	Q03_NN_Male.wav	57.59
7	Q4.wav	Q04_Eric.wav	298.23
8	Q4.wav	Q04_NN_Male.wav	279.03
9	Q5.wav	Q05_Eric.wav	25.59
10	Q5.wav	Q05_NN_Male.wav	15.86
11	Q6.wav	Q06_Eric.wav	132.09
12	Q6.wav	Q06_NN_Male.wav	88.57
13	Q7.wav	Q07_Eric.wav	10.23
14	Q7.wav	Q07_NN_Male.wav	6.39
15	Q8.wav	Q08_Eric.wav	26.62
16	Q8.wav	Q08_NN_Male.wav	15.86
17	Q9.wav	Q09_Eric.wav	32.76
18	Q9.wav	Q09_NN_Male.wav	16.89
19	Q10.wav	Q10_Eric.wav	33.53
20	Q10.wav	Q10_NN_Male.wav	18.68
21	R1.wav	R01_Jos.wav	109.01
22	R1.wav	R01_Peter.wav	432.29
23	R2.wav	R02_Jos.wav	44.79
24	R2.wav	R02_Peter.wav	197.62

- Between-sources variability estimation: Comparison between the features of the questioned recording and the statistical models of the voices of the speakers from the database representing the potential population.
- Calculation of the evidence score: Comparison between the questioned recording and the model of the suspected speaker.
- Calculation of the strength of evidence: Calculation of the likelihood ratio by evaluating the relative likelihood ($\frac{p(E|H_0)}{p(E|H_1)}$) of observing the evidence score (E) given the hypothesis that the source of the questioned recording is the suspect (H_0) and the likelihood of observing the evidence score given hypothesis that someone else in the potential population was its source (H_1) Kernel density estimation was used to calculate the probability densities of distribution of scores for each of the hypotheses.

4.6 Results of the Analysis

Each of the ten questioned recordings ($Q1, ..., Q10$) is considered as a separate case. For each case, we consider the question, 'Is the speaker Peter, in the reference recordings $R1$ and $R2$, the same speaker as the unknown speaker in the questioned recording (Qn)?'

The following databases are used in each case:

- Potential Population Database (P): PolyCOST 250 Database,
- Reference Database (R): 2 reference recordings from the recordings $R1$ and $R2$, belonging to the suspected speaker, Peter
- Control Database (C): 7 control recordings from the recordings $R1$ and $R2$ belonging to the suspected speaker.

The distribution of scores for H_0 obtained when comparing the features of the suspected speaker control recordings (C database) of the suspected speaker, Peter, with the two statistical models of his speech (created using files from the R database) is represented by the dotted line. In this paper, an example analysis is done for the Case 3.

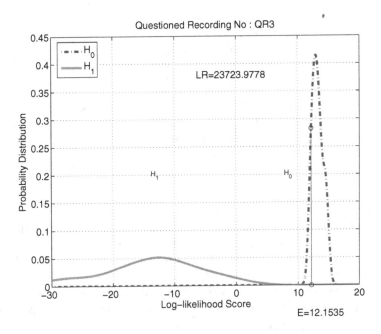

Fig. 2. *Case 3: Questioned recording Q03_NN_Male.wav*

Case 3

Is Peter in the reference recordings ($R1$ and $R2$) the same speaker as the unknown speaker in the recording $Q3$?

Trace Database (T): Q03_NN_Male.wav containing the speech of the unknown speaker from the questioned recording $Q3$.

Discussion. The distribution of scores for H_1 obtained by comparing the segment of the questioned recording $Q3$, corresponding to the unknown speaker (Q03_NN_Male), with the Gaussian mixture models of the speakers of the potential population database (P) is represented by the solid line in Fig. 2.

The average score (E), represented by the point on the log-likelihood score axis in Fig. 2, obtained by comparing the questioned recording with the Gaussian mixture models of the suspected speaker, Peter's speech is 12.15.

A likelihood ratio of 23723.98, obtained in Fig. 2, means that it is 23723.98 times more likely to observe this score (E) given the hypothesis H_0 (the suspect is the source of the questioned recording) than given the hypothesis H_1 (that another speaker from the relevant population is the source of the questioned recording).

We also observe that this score of E, is statistically significant (at a 5% statistical significance level) in the distribution of scores corresponding to hypothesis H_0.

All Cases

A summary of the results of the analyses is presented below in Table 2.

Table 2. LRs obtained using the PolyCOST 250 database, in English, as the potential population

Trace No.	Biometric Evidence	LR (P in English)	Correct
Q1	10.86	6.56	Inconclusive
Q2	11.20	163.41	Inconclusive
Q3	12.15	23723.98	✓
Q4	12.68	21720.97	✓
Q5	13.51	11631.8	✓
Q6	11.63	329.0	✓
Q7	12.48	38407.33	Rejected
Q8	10.68	0.660	Inconclusive
Q9	12.92	3033.47	✓
Q10	7.19	4.36 x 10-23	Inconclusive

The conclusions with respect to each of the ten questioned files have in each case been placed on the scale of conclusions that the expert uses. In Table 2, they are designated as correct or incorrect (the strength of the biometric evidence (E) is given by LR), inconclusive, or rejected. The latter category includes the results for files that are judged to be too short $(Q7)$, and the category inconclusive includes results that are not statistically significant $(Q1, Q2, Q8, Q10)$. This means that the statistical significance analysis does not allow us to progress the case in any direction. The conclusions of the remaining participants of the ENFSI-NFI speaker recognition evaluation through a fake case are presented in [7] for comparison.

5 Conclusions

The main objective of this paper is to show what can be done in the domain of forensic automatic speaker recognition using statistical evaluation of biometric evidence. This paper gives step-by-step guidelines for the calculation of the biometric evidence and its strength under operating conditions of the casework. In the paper, an automatic method using the Gaussian mixture models (GMMs) and the Bayesian interpretation (BI) framework were implemented in the forensic speaker recognition task.

This paper also reports on the first ENFSI evaluation campaign through a fake case, organized by the Netherlands Forensic Institute (NFI), as an example, where the proposed automatic method was applied.

The aim of the analysis was to determine whether the recordings of unknown speakers, in the ten questioned recordings, were produced by the suspected speaker, named Peter in the reference recordings. Note that the given conclusions take into consideration the likelihood ratios as well as other factors such the length and content of the recordings and the statistical significance of the results. These factors may influence the statement of the conclusions from the likelihood ratio or do not allow us to progress the case in any direction.

References

1. Jain, A., et al. (eds.): Handbook of Biometrics. Springer, New York (2008)
2. Rose, P.: Forensic Speaker Identification. Taylor and Francis, London (2002)
3. Drygajlo, A.: Forensic Automatic Speaker Recognition. IEEE Signal Processing Magazine 24(2), 132–135 (2007)
4. Aitken, C., Taroni, F.: Statistics and the Evaluation of Evidence for Forensic Scientists. John Wiley and Sons, Chichester (2004)
5. Champod, C., Meuwly, D.: The Inference of Identity in Forensic Speaker Identification. Speech Communication 31(2-3), 193–203 (2000)
6. Drygajlo, A., Meuwly, D., Alexander, A.: Statistical Methods and Bayesian Interpretation of Evidence in Forensic Automatic Speaker Recognition. In: Proceedings of 8th European Conference on Speech Communication and Technology (Eurospeech 2003), Geneva, Switzerland, pp. 689–692 (2003)
7. Cambier-Langeveld, T.: Current methods in forensic speaker identification: Results of a collaborative exercise. The International Journal of Speech, Language and the Law 14.2, 223–243 (2007)
8. Alexander, A., Drygajlo, A., Botti, F.: NFI: Speaker recognition evaluation through a fake case. Case Report, EPFL, Lausanne (2005)

Forensic Authorship Attribution Using Compression Distances to Prototypes

Maarten Lambers[2] and Cor J. Veenman[1,2]

[1] Intelligent Systems Lab,
University of Amsterdam, Amsterdam, The Netherlands
[2] Digital Technology & Biometrics Department,
Netherlands Forensic Institute, The Hague, The Netherlands

Abstract. In several situations authors prefer to hide their identity. In forensic applications, one can think of extortion and threats in emails and forum messages. These types of messages can easily be adjusted, such that meta data referring to names and addresses is at least unreliable. In this paper, we propose a method to identify authors of short informal messages solely based on the text content. The method uses compression distances between texts as features. Using these features a supervised classifier is learned on a training set of known authors. For the experiments, we prepared a dataset from Dutch newsgroup texts. We compared several state-of-the-art methods to our proposed method for the identification of messages from up to 50 authors. Our method clearly outperformed the other methods. In 65% of the cases the author could be correctly identified, while in 88% of the cases the true author was in the top 5 of the produced ranked list.

1 Introduction

Besides the many legitimate uses of the internet, it also has become a communication medium for illegal activities. Examples include child pornography distribution, threatening letters and terroristic communications. Finding the illegal distributor or the writer of illegal content is an important aspect of the efforts of the forensic experts to reduce criminal activity on the internet.

In this research, we focus on the analysis of short messages such as e-mail, newsgroup and forum messages. To determine the source of such a communication item, the most straightforward approach seems to use the addressee from the meta data of the message. Unfortunately this information is very unreliable, because this type of data can be changed easily at will. Moreover, even if the addressee was an email address of the true sender, the address and identity can still be meaningless. Obtaining an arbitrary email account name from a webmail provider can be done easily and anonymously.

Another option is to focus on network traffic, i.e. to find out how a message was routed or to monitor the IP addresses e.g. of the users accessing a terrorist website [1]. While this can provide important clues in headers and meta data, there are also several ways to disguise the source of a packet, i.e. message. Another downside of this approach is that only a link from a message to a computer can be established. For forensic purposes it is more useful to attribute the authorship of a message, that is to link the writings to the actual writer.

Z.J.M.H. Geradts, K.Y. Franke, and C.J. Veenman (Eds.): IWCF 2009, LNCS 5718, pp. 13–24, 2009.

Most of the early research in the field of authorship attribution deals with the disputed authorship of papers or books [2], the best known example being the case of the Federalist Papers (see for example [3] and [4]). In the forensic domain, where texts are typically relatively short and the number of potential authors relatively high, different methods are required [5], [6], [7], [8], [9]. Expectedly, the reported performances in these works are dependent on the number of authors used in the experiments [10]. The model from [9] has an accuracy of 75% on a 20-class problem, whereas an accuracy of 90% is reported in [5] on a 5-class problem. From [8] can be learned that an author is more difficult to categorize compared to other authors when fewer texts are available of that particular author. In [7] it is concluded that 20 messages containing approximately 100 words are sufficient for authorship attribution. However, these results are based on experiments conducted on e-mail messages from only four different authors. Accordingly, the researches mentioned are not yet useful for forensic applications, considering the accuracies and the constraints; either the accuracy is too low (30% in [6]) or the number of classes is too low (three to five in [5], [8] and [7]). An appropriate method must be able to handle tens of different authors and must have a higher accuracy.

In this paper, we deal with a scenario that is more realistic from a forensic perspective. We propose a method for author attribution from a set of 50 known authors, which is a lot more than in previously reported research. Such a scenario fits a situation were a substantial though limited list of targets is followed for tactical investigations. As [11], [12], [13], the proposed method is based on compression distances between texts. To cope with the computational as well as modelling problems that typically arise with these instance based approaches [10], we propose to use a set of texts as prototypes to which the compression distances are computed. These distances serve as feature representation which allows for the training of supervised classification models. We further use ranked lists as output of the models to fit the investigation scenario and accommodate for some classification errors. That is, in case of classification error the target is likely one of the following candidates on the ranked list.

The next section, Section 2, formally defines the problem, where the constraints of this research are elaborated. This is followed by an overview of related work on authorship attribution in Section 3. In Section 4, we describe our proposed method. The experiments are described subsequently in Section 5, including data description, feature extraction and the results. A discussion of the results with pointers for additional work concludes this paper in Section 6.

2   Problem Statement

The goal of this research is to assign an author's identity to a text message without the use of any meta data. The list of potential authors is considered known. This is known as a closed set recognition problem [14].

Further, we constrain the research to short and informal messages. E-mails as well as newsgroup or forum messages are considered applicable data sources. These examples can be seen as similar data considering the informality and use of such messages. However, the meta data and the text structure can be different. That is, features based on attachments (as in [8]) are not applicable on forum messages and features based on fonts (the size or the colour of the font, as in [5]) might not be applicable to e-mails. This constraint restricts the problem to basic text content (UTF-8).

For investigation purposes, typically the required output is a ranked list. Therefore, also for the author attribution problem the output should be a ranked list. That is, ranked in order of relevance, where the most probable author is ranked first.

3 Text Features

In order to learn a classification model for authorship attribution, we need a representation of the texts that enables to differentiate one author from the other. Traditionally, the objects, here texts, are represented using a set of properly chosen features. Approximately 1,000 features or style markers are known in the field of authorship attribution that have proved to be successful in certain scenarios [15]. We describe them shortly and focus on those features that we use in our experiments. We follow the feature categories as proposed in [10]. Slightly different categorizations have been proposed elsewhere [16], [17], [9].

3.1 Lexical Features

Lexical features describe a range of ratios, averages, lengths, counts and frequencies of words, lines and characters. They are supposed to profile the author's vocabulary richness computed as the ratio between the number of different words and the total number of words [18], statistics about the sentences as average length of sentences, number of sentences [9], and relative frequencies of special characters (punctuation characters, (tab) spaces, digits and upper- and lowercase characters) [19].

3.2 Syntactic Features

Syntactic features refer to the patterns used for the formation of sentences, using part-of-speech tagging and function words. These features are considered beyond an author's conscious control, for instance short all-purpose words [12]. The author thus leaves a writeprint [17] in his or her text by using certain words in certain frequencies. The term function words is referred to by many papers in the field of text mining: [16], [6], [20], [7], [8], [21], [22], [17], [4], [23], [24], [9]. This term describes words that do not contain information about the document's content such as prepositions, pronouns or determiners [22] or can be defined as words that are context-independent and hence unlikely to be biased towards specific topics [25]. Previous studies show the value of features based on these function words. In [7], they are referred to as the best single discriminating feature. The number of function words used in experiments reported range from 50 words [20] to 365 words [24].

3.3 Semantic Features

Using Natural Language Technology (NLP) additional information about texts can be obtained, such as part-of-speech tags. This can give insight in the use of different classes of words (noun, verb, adjective, etc.). Having semantic knowledge about the text can also help identifying synonyms, hypernyms or antonyms, thus gaining knowledge about the context of the text on which features can be based.

3.4 Application Specific Features

Application specific features contain structural features and content-specific features. Structural features are based on the structure and layout of a text. This includes information about indentation, paragraph lengths and the use of a greeting statement [9], or the font size or font colour, the use of links and the use of images [5]. Content-specific features denote the use of certain key words belonging to the subject of the texts. This can be of use when all the available texts discuss the same topic. Carefully selected content-based information may reveal some authorial choices [10].

3.5 Character Features

The use of character n-grams in the field of authorship attribution is proposed in [7] and [26]. Character n-grams are simply substrings of the original text. When $n = 2$, the word 'text' results in the character bi-grams 'te', 'ex' and 'xt'. The features generated from these bi-grams are typically the relative frequency of occurrence within a text. Some experiments have used only the 26 characters of the alphabet [26] while others also use punctuation characters [7].

In the research described in [26], a study in authorship attribution is conducted where only features based on character n-grams are used. The goal of the investigation was to find the optimal number n. The characters that are used in [26] only consist of the 26 character of the alphabet. The experiments that are conducted let n run from 1 to 10. The other variable in these experiments is the profile size L, denoting the number of features. With character 1-grams, the number of dimensions representing the object is only 26. With every increase of variable n, the number of dimensions multiplies with 26, making character 3-gram features already computationally very complex. The profile size L limits the number of dimensions by using only the L most frequently used n-grams. Generally, more dimensions results in a higher accuracy according to these experiments. However, the complexity of the problem reduces the accuracy, when the number of dimensions becomes too high. Experiments on English data suggest the use of 5 to 7-grams, while experiments on Greek data slightly differ. When considering a profile size of 500, which seems a reasonable amount of dimensions, 2 or 3-grams work best [26], [27].

Compression-based Approaches. Another type of methods categorized as exploiting character features are compression-based approaches [11], [28]. They can be considered character features, because compression models also describe text characteristics based on repetition of character sequences. The basic idea is that texts that have similar characteristics should be compressed with each others compression model (dictionary) easily. That is, without the need for substantial extension of the compression model (dictionary).

4 Compression Distances to Prototypes (CDP)

Compression-based similarity methods are motivated from the Kolmogorov complexity theory [29]. They have recently gained popularity in various domains, ranging from the construction of language trees [30], the phylogenetic tree [31] and plagiarism detection [32]. The compression-based approaches are practical implementations of the information distances expressed in the non-computable Kolmogorov complexity. Several compression distances have been defined [11], [33], [28], [13], [34], [35].

We follow the Compression Distance Metric (CDM) defined in [13]:

$$CDM(\mathbf{x}, \mathbf{y}) = \frac{C(\mathbf{xy})}{C(\mathbf{x}) + C(\mathbf{y})}, \tag{1}$$

where $C(\mathbf{x})$ is the size of the compressed object $\mathbf{x}$ and $\mathbf{xy}$ is the concatenation of $\mathbf{x}$ and $\mathbf{y}$.

The CDM expresses the compression distance between two objects $\mathbf{x}$ and $\mathbf{y}$ as the difference in compressed length of the concatenation of the two objects $\mathbf{xy}$ divided by the sum of the length of the individually compressed objects, which can be computed easily using various compressors. If objects $\mathbf{x}$ and $\mathbf{y}$ are similar, then $CDM(\mathbf{x}, \mathbf{y}) < 1$. The more similar $\mathbf{x}$ and $\mathbf{y}$ are, the smaller the CDM value becomes. While it approaches 0.5, when $\mathbf{x}$ and $\mathbf{y}$ are equal. On the other hand the CDM will approach 1 for objects that are completely unrelated [13]. Clearly, these properties of CDM depend on the quality of the compressor.

Having a distance metric to find similar objects, a straightforward approach to recognize unseen objects is to attribute them to the author of the most similar object in the training database. In other words, the nearest neighbour classification method can be applied. This *instance-based approach* [10] is followed by [11]. There are, however, some downsides to the use of the nearest neighbour rule. When using this approach, to classify an unseen object $\mathbf{x}$, the distances between $\mathbf{x}$ and all training objects need to be calculated. This is a computationally expensive procedure. Another problem is that the nearest neighbour approach runs the risk of overfitting [36]. Alternatively, all training texts from the same category, here an author, can be merged. In such a *profile-based approach* [10], the compression distance is computed between the unseen text and the merged author corpus. Unseen texts are then attributed to the closest author corpus [12]. This can be seen as converting all text prototypes into one, similar to the nearest mean classifier approach.

We propose another way to reduce the computational load and the overfitting. We construct a feature vector from the distances to the training texts, a dissimilarity-based classification approach [37]. As a result, the feature vector could become as long as the number of objects in the training set. This poses, however, new computational and learning problems. Still, the distances have to be computed from a test text to all training database texts. Moreover, the resulting enormous feature vectors make the learning of models costly. More importantly, the feature space will be too sparsely sampled leading to dimensionality curse problems. While dimension reduction techniques and robust classifiers could be considered, that is deemed outside the scope of this work.

Instead, in our approach we first select a subset of *prototypes* of the training set. Then, for each training text, the CDM to all prototypes is computed. Accordingly, we obtain a feature vector consisting of distances to the prototypes. The idea of using a subset as prototypes is that such a subset of the data can describe the training set sufficiently [38]. It should be noted that for dissimilarity-based feature vectors, prototype selection is the same as feature selection.

Once a feature vector is computed with distances to the prototypes, any feature based classifier can be applied. Since our problem is author attribution for a closed set of authors, classifiers that are applicable to multi-class problems are preferred. Though also combining schemes can be considered, such as one-versus-all and one-versus-one.

5   Experiments

In order to test the relative performance of the proposed Compression Distance to Prototypes (CDP) method, we need to compare it to other methods known in the literature. The related methods that we consider in the comparison are based on feature sets as described in Section 3 and further specified in this Section. Besides a feature representation, a classifier has to be chosen. Based on unreported experiments, we have selected the Fisher discriminant classifier (e.g. [39]) for all methods and experiments. Its performance is competitive, while it has also computational attractive properties. Below, we first describe the experimental dataset and the required preprocessing. Then, we elaborate on the compared feature sets. We also describe additional details that need specification for practical application of the proposed method. We further describe the evaluation protocol and the actual experiments.

5.1   Dataset and Preparation

The data we have used in the experiments originates from Dutch newsgroups. They are mostly informally written and are varying in length. We collected the data from four newsgroups, covering the subjects of politics, jurisdiction, motorbikes and cars. These newsgroups are chosen because the contributors to these newsgroups are assumed to discuss serious subjects and have no intention to fake their identities, which makes the author labels reliable.

We selected messages of authors that have sent at least 100 messages. This resulted in a dataset with 130 authors with in total 55,000 messages.

Fig. 1(a) shows the distribution of authors (the vertical axis) that have written a certain number of messages. As can be seen in the figure, most authors have written around 300 messages and only few persons have written more than 1000 messages. The length of these messages is dispersed in a similar fashion: the bulk of messages are short (88% of the messages have less than 100 words), see Fig. 1(b).

The newsgroup messages that have been used in the experiments have been prepared in order to remove headers, quoted text, greetings and signatures. No text is allowed to have a direct link to the author; the resulting text should only consist of the *body* of the message. Signatures are replaced by the string 'signature', any the name or nickname of the author is replaced by respectively 'Firstname', 'Surname' or 'Nick'. This is chosen over deleting this information in order to be able to distinguish between authors who do use a signature/greeting and authors who do not.

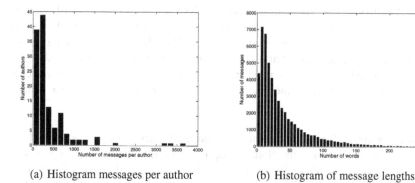

(a) Histogram messages per author   (b) Histogram of message lengths

Fig. 1. Statistics of the dataset

Randomly sampling the prepared messages confirms the absence of headers etc., although some other unwanted text is sometimes included in the resulting bodies. Quotes from previous messages on the newsgroup are preceded by characters such as '>', but when a user pastes a newspaper article or law book excerpt in their message, this text is not automatically recognized. In roughly three to four messages in every 100, pasted text is found. This makes the authorship attribution problem a bit more difficult, but checking every message by hand is not feasible.

5.2   Evaluation Protocol

For the performance evaluation of the different methods, we have used 10-fold cross-validation repeated 20 times. The reported performance is the fraction of messages for which the author is correctly classified. That is, the messages for which the correct author is ranked first in the yielded ranked list (the required output). We also use the Cumulative Match Characteristic (CMC) Curve to report system performance [14]. This curve plots the rank against the fraction of messages recognized below and including that rank.

5.3   Compared Feature Sets

We created three feature sets to compare the proposed method to. These feature sets are based on features reported in the literature and fit reasonably to the taxonomy as described in Section 3.

Lexical Features. We based our lexical features on [19] and [9] that both used features that measure the vocabulary richness from [18]. These features are based on the number of different words and the total number of words in a text. Furthermore the lexical feature set contains the frequency of punctuation characters, the word length histogram ranging from 1 to 30 characters which delivers 30 features, and the average sentence and word length.

Syntactic Features. The syntactic features that we used are function word features. Some papers list a predefined set of function words [40], [9]. Others simply use the most frequently used words as function words [4], [18]. The first approach requires a validated word list per language. This makes the latter approach more generic. Although it may seem that mainly context-dependent words are found as most frequent, the contrary has been found in [20]. Since this is also our experience, we used this generic approach.

Character Features. We used character 2-grams as character features as in [7], [27], and [26]. The 26 characters of the alphabet together with the space character generates a total of $27 \cdot 27 = 729$ features. We did not include punctuation characters. The feature value is the number of occurrences of the respective character combination divided by the total number of characters in the message. There has been no selection of the most discriminating bi-grams, as was exercised in [26].

5.4   Parameters and Tuning

For the syntactic features, the optimal number of function words can be tuned. For the tuning we sampled datasets with the messages of 50 authors, 50 messages per author and a minimum message length of 50 words. In Fig. 2(a), the cross-validated average classification performance is displayed as a function of the number of function words. For this problem setting the optimum number of function words is around 600.

For the CDP method, there are two parameters to settle in order to obtain a compression-based set of distances to prototypes. First, a proper compressor has to be found. The most suitable compressor is probably domain dependent. In some studies the RAR scheme [41] has been found a good general purpose candidate [42], [34]. For practical implementation reasons we applied the LZ76 compression method [43], [44]. The other important parameter is the number prototypes and the way they are selected. Various prototype selection schemes have been researched and compared [45], [46]. Because random sampling is competitive in performance to advanced and computationally involved methods [45], [46], we applied random prototype selection.

For the tuning of the number of prototypes, we used the same set-up as for the function words tuning, that is, samples of 50 authors, 50 messages per author and a minimum message length of 50 words. Fig. 2(b) shows the cross-validated average classification performance. With more than fifteen prototypes the performance is maximal. Consequently, fifteen is the most suitable number of prototypes for this problem setting. In a similar fashion the number of prototypes is determined when 50 authors all contribute 90 messages of at least 50 words in length.

5.5   Results

Together with the three feature sets as described above, the proposed CDP method is put to the test. In a direct competition, our method shows a significantly better result compared to the other feature sets, of which the lexical features rank second best, see Fig. 3(a). The difference in performance is very noticeable: an authorship attribution system using the lexical features correctly identifies the true author in 40% of the cases,

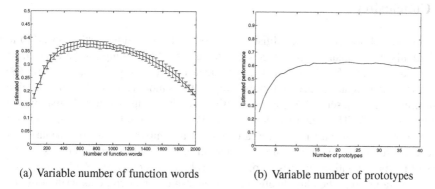

(a) Variable number of function words   (b) Variable number of prototypes

Fig. 2. Parameter tuning graphs with average performance and standard error bars

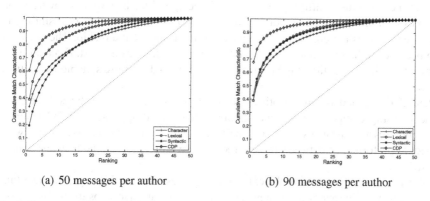

(a) 50 messages per author   (b) 90 messages per author

Fig. 3. Systems performance displayed as Cumulative Match Characteristic (CMC) Curves. The messages of 50 different authors, contributing messages of at least 50 words in length, are used for these experiments. (The standard deviation is left out for readability purposes.)

whereas a similar system using compression distances to prototypes scores 60%. Bear in mind that given the high number of authors, a random classifier would only correctly identify 2%.

When the resulting top five is inspected, one can see that in 83% of the cases the true author of a questioned document is found in the five most likely authors using the CDP method. This in contrast to 70% using a system with the second best set of features. The steepness of the CMC curve gradually decreases and a certainty nearing the 100% is only found when the first 30 of the 50 ranked authors are inspected.

The same experiment is repeated with an increase of messages per author from 50 to 90. Fig. 3(b)) shows similar results; again the proposed CDP method outperforms the known methods. As expected, the system scores better when more messages per author are available for training.

6 Conclusion

This research shows that the established methods for authorship attribution perform significantly worse than our proposed CDP feature set. Depending on the dataset, an increase around 20-25% in accuracy can be measured (see Fig. 3(a) and 3(b)). The performance gain makes practical forensic applications more feasible. Given a suitable set of training data, a set of 50 suspected authors can be comprised to only five, with a certainty near 90%. This heavily reduces the workload. It should be noted that a closed set situation, as dealt with in this research, is not a completely realistic scenario for tactical investigators. An interesting continuation of this research could be to shift the focus of the investigation towards an open set problem. An open set problem does not give the certainty that the classification model is trained on documents written by the author of a questioned document, as this research has. An ideal model would then be able to distinguish between the text of an unknown author and the texts from the dataset it has trained on, thus identifying the new text as being written by a yet unknown author.

An extension towards an authorship verification model is then also a possibility. In this case two texts are compared, of which one is of known authorship. The question then becomes if the second text is written by the same author as the first. From a forensic point-of-view, an authorship verification model would be more suitable for tactical purposes.

The CDP method is among others proposed as alternative to instance-based approaches using compression distances. Like profile-based approaches, CDP features make compression distances computationally feasible. A experimental comparison with respect to performance between these three compression-based methods is also worth investigating.

Considering the data used in the experiments as described in this paper, the fact that messages from only four different newsgroups are used can have a positive influence on the results. Although authors were found contributing to discussions in multiple groups, certain authors tend to stick to certain newsgroups. This could work to the advantage of the classification method, although it is likely that it would not favour any feature set in particular. As a recommendation for future work, similar experiments could be conducted on a more diverse set of messages and authors. Another recommendation for future work is to investigate the proposed method on messages written in another language (e.g. English).

References

1. Elovici, Y., Kandel, A., Last, M., Shapira, B., Zaafrany, O.: Using data mining techniques for detecting terror-related activities on the web. Journal of Information Warfare 3(1), 17–29 (2004)
2. Stamatatos, E., Fakotakis, N., Kokkinakis, G.: Computer-based authorship attribution without lexical measures. Computers and the Humanities 35(2), 193–214 (2001)
3. Holmes, D., Forsyth, R.: The federalist revisited: New directions in authorship attribution. Literary and LInguistic Computing 10(2), 111–127 (1995)
4. Mosteller, F., Wallace, D.: Inference and disputed authorship: the Federalist. Addison-Wesley, Reading (1964)

5. Abbasi, A., Chen, H.: Applying authorship analysis to extremist-group web forum messages. IEEE Intelligent Systems 20(5), 67–75 (2005)
6. Argamon, S., Sari, M., Stein, S.: Style mining of electronic messages for multiple authorship discrimination: first results. In: Proceedings of the 9th ACM SIGKDD, pp. 475–480 (2003)
7. Corney, M., Anderson, A., Mohay, G., Vel, O.D.: Identifying the authors of suspect e-mail. Technical report, Queensland University of technology (2001)
8. Vel, O.D., Anderson, A., Corney, M., Mohay, G.: Mining e-mail content for author identification forensics. ACM SIGMOD Record 30(4), 55–64 (2001)
9. Zheng, R., Li, J., Chen, H., Huang, Z.: A framework of authorship identification for online messages: Writing style features and classification techniques. Journal American Society for Information Science and Technology 57(3), 378–393 (2006)
10. Stamatatos, E.: A survey of modern authorship attribution methods. Journal of the American Society for Information Science and Technology 60(3), 538–556 (2009)
11. Benedetto, D., Caglioti, E., Loreto, V.: Language trees and zipping. Phys. Rev. Lett. 88(4), 048702 (2002)
12. Khmelev, D.V., Teahan, W.: A repetition based measure for verification of text collections and for text categorization. In: 26th Annual International ACM SIGIR Conference on Research and Development in Information Retrieval, Toronto, Canada, August 2003, pp. 104–110 (2003)
13. Keogh, E., Lonardi, S., Ratanamahatana, C.: Towards parameter-free data mining. In: Proceedings of the 10th ACM SIGKDD International Conference on Knowledge Discovery and Data Mining, pp. 206–215 (2004)
14. Bolle, R., Connell, J., Pankanti, S., Ratha, N., Senior, A.: Guide to Biometrics. Springer, New York (2004)
15. Rudman, J.: The state of authorship attribution studies: Some problems and solutions. Computers and the Humanities 31(4), 351–365 (1998)
16. Abbasi, A., Chen, H.: Visualizing authorship for identification. In: Mehrotra, S., Zeng, D.D., Chen, H., Thuraisingham, B., Wang, F.-Y. (eds.) ISI 2006. LNCS, vol. 3975, pp. 60–71. Springer, Heidelberg (2006)
17. Li, J., Zeng, R., Chen, H.: From fingerprint to writeprint. Communications of the ACM 49(4), 76–82 (2006)
18. Tweedie, F., Baayen, R.: How variable may a constant be? measure of lexical richness in perspectiv. Computers and the Humanities 32(5), 323–352 (1998)
19. Corney, M.: Analysing e-mail text authorship for forensic purposes. Master's thesis, Queensland University of technology (2003)
20. Binongo, J.: Who wrote the 15th book of oz? an application of multivariate analysis to authorship attribution. Chance 16(2), 9–17 (2003)
21. Gamon, M.: Linguistic correlates of style: authorship classification with deep linguistic analysis features. In: Proceedings of the 20th International Conference on Computational Linguistics, pp. 611–617 (2004)
22. Kaster, A., Siersdorfer, S., Weikum, G.: Combining text and linguistic document representations for authorship attribution. In: In SIGIR Workshop: Stylistic Analysis of Text for Information Access, pp. 27–35 (2005)
23. Uzuner, O., Katz, B.: A comparative study of language models for book and author recognition. In: Dale, R., Wong, K.-F., Su, J., Kwong, O.Y. (eds.) IJCNLP 2005. LNCS, vol. 3651, pp. 969–980. Springer, Heidelberg (2005)
24. Zhao, Y., Zobel, J.: Effective and scalable authorship attribution using function words. In: Lee, G.G., Yamada, A., Meng, H., Myaeng, S.-H. (eds.) AIRS 2005. LNCS, vol. 3689, pp. 174–189. Springer, Heidelberg (2005)

25. Koppel, M., Schler, J.: Exploiting stylistic idiosyncrasies for authorship attribution. In: Proceedings of IJCAI 2003 Workshop on Computational Approaches to Style Analysis and Synthesis, pp. 69–72 (2003)
26. Kešelj, V., Peng, F., Cercone, N., Thomas, C.: N-gram-based author profiles for authorship attribution. In: Proceedings of the Conference Pacific Association for Computational Linguistics, pp. 255–264 (2003)
27. Kjell, B.: Authorship attribution of text samples using neural networks and bayesian classifiers. IEEE International Conference on Systems, Man and Cybernetics 2, 1660–1664 (1994)
28. Li, M., Chen, X., Li, X., Ma, B., Vitányi, P.M.B.: The similarity metric. IEEE Transactions on Information Theory 50(12), 3250–3264 (2004)
29. Li, M., Vitányi, P.M.B.: An Introduction to Kolmogorov Complexity and its Applications. Springer, New York (1997)
30. Ball, P.: Algorithm makes tongue tree. Nature (January 2002)
31. Li, M., Badger, J.H., Chen, X., Kwong, S., Kearny, P., Zhang, H.: An information-based sequence distance and its application to whole mitochondrial genome phylogeny. Bioinformatics 17(2), 149–154 (2001)
32. Chen, X., Francia, B., Li, M., McKinnon, B., Seker, A.: Shared information and program plagiarism detection. IEEE Transactions on Information Theory 50(7), 1545–1551 (2004)
33. Cilibrasi, R., Vitányi, P.M.B.: Clustering by compression. IEEE Transactions on Information Theory 51(4), 1523–1545 (2005)
34. Kukushkina, O.V., Polikarpov, A.A., Khmelev, D.V.: Using literal and grammatical statistics for authorship attribution. Problems of Information Transmission 37(2), 172–184 (2001)
35. Telles, G., Minghim, R., Paulovich, F.: Normalized compression distance for visual analysis of document collections. Computers and Graphics 31(3), 327–337 (2007)
36. Mitchell, T.: Machine Learning. McGraw-Hill, New York (1997)
37. Pękalska, E., Skurichina, M., Duin, R.: Combining fisher linear discriminants for dissimilarity representations. In: Kittler, J., Roli, F. (eds.) MCS 2000. LNCS, vol. 1857, pp. 117–126. Springer, Heidelberg (2000)
38. Duin, R., Pękalska, E., Ridder, D.D.: Relational discriminant analysis. Pattern Recognition Letters 20(11–13), 1175–1181 (1999)
39. Duda, R., Hart, P., Stork, D.: Pattern Classification. John Wiley and Sons, Inc, New York (2001)
40. Craig, H.: Authorial attribution and computational stylistics: If you can tell authors apart, have you learned anything about them? Literary and Linguistic Computing 14(1), 103–113 (1999)
41. Roshal, E.: RAR Compression Tool by RAR Labs, Inc (1993-2004), http://www.rarlab.com
42. Marton, Y., Wu, N., Hellerstein, L.: On compression-based text classification. In: Advances in Information Retrieval, pp. 300–314 (2005)
43. Lempel, A., Ziv, J.: On the complexity of finite sequences. IEEE Transaction on Information Theory 22, 75–81 (1976)
44. Kaspar, F., Schuster, H.: Easily calculable measure for the complexity of spatiotemporal patterns. Physical Review A 36(2), 842–848 (1987)
45. Kim, S.W., Oommen, B.J.: On using prototype reduction schemes to optimize dissimilarity-based classification. Pattern Recognition 40(11), 2946–2957 (2007)
46. Pekalska, E., Duin, R., Paclik, P.: Prototype selection for dissimilarity-based classifiers. Pattern Recognition 39, 189–208 (2006)

Estimation of Inkjet Printer Spur Gear Teeth Number from Pitch Data String of Limited Length

Yoshinori Akao[*], Atsushi Yamamoto, and Yoshiyasu Higashikawa

National Research Institute of Police Science, Kashiwa, Chiba 2770882, Japan
akao@nrips.go.jp

Abstract. In this paper, we investigate the feasibility of estimating the number of inkjet printer spur gear teeth from pitch data strings of limited length by maximum entropy spectral analysis. The purpose of this study is to improve the efficiency of inkjet printer model identification based on spur mark comparison in the field of forensic document analysis. Experiments were performed using two spur gears in different color inkjet printer models, and eight different lengths of pitch data strings—ranging from three to 10 times the number of spur gear teeth. The result for a data string longer than five times the number of teeth showed a proper estimation within a deviation of one tooth. However, the estimation failed for shorter data strings because the order in maximum entropy analysis was determined inappropriately. The presented results provide information on the number of spur gear teeth from shorter data strings than in a previous study.

Keywords: forensic document analysis, inkjet printer identification, spur mark comparison method, pitch, spectral analysis, maximum entropy method.

1 Introduction

Computational approaches have provided extremely effective solutions for problems in forensic document analysis [1, 2, 3]. However, various fraudulent documents, such as counterfeited banknotes, securities, passports, traveler's checks and driver's licenses, are still being created worldwide on computers and peripherals as a result of technological improvements. These technologies are also used to counterfeit documents such as wills, contracts, and receipts that can be used to perpetrate financial crimes.

Inkjet printer identification is becoming more prevalent in document analysis, and of the various computer peripherals, the number of inkjet printers is drastically increasing and it is becoming the most popular image output device for personal use [4]. In the first stage of forensic studies, the chemical or spectral properties of inkjet inks are the main focus [5, 6, 7].

Tool marks have been extensively used in forensic science to link a known and a suspect item, and the items that created the markings [8]. Also in the field of forensic document analysis, examinations based on tool marks have been applied to photocopiers

---

[*] Corresponding author.

Z.J.M.H. Geradts, K.Y. Franke, and C.J. Veenman (Eds.): IWCF 2009, LNCS 5718, pp. 25–32, 2009.

[9, 10, 11], typewriters [12, 13, 14], and label markers [15]. The latent physical markings left on documents by printers and copiers have also been visualized by electrostatic detection devices [16].

Based on the philosophy of tool mark examination, the authors present a spur mark comparison method (SCM) [17, 18, 19], in which we compare the spur gears of an inkjet printer with the spur marks on printouts. Spur gears are mechanisms that hold the paper in place as it passes through an inkjet printer [20], and spur marks are the indentations left on documents by the spur gears. In order to improve the feasibility of SCM, Furukawa *et al.* proposed a visualization method for spur marks using an infrared (IR) image scanner, and enhancing image quality by estimating a point spread function of the scanner [21]. In SCM, the "pitch" and "mutual distance" indices are introduced to classify inkjet printer models. The greater the number of indexes, the more precise the classification.

In our previous study [22], we extracted data on the number of spur gear teeth from data strings. There is a periodicity in spur mark pitch data strings, as spur marks are formed by the same part of a spur gear at regular tooth number intervals. The teeth number of spur gears was successfully estimated within a deviation of one tooth. However, the estimation was only performed for pitch data string with a length of 10 times the number of teeth, therefore the applicability of the proposed method for shorter data strings was not proved.

In this paper, we investigate the feasibility of estimating the number of inkjet printer spur gear teeth from pitch data strings of limited length, using a computational approach—maximum entropy spectral analysis.

2 Materials and Method

2.1 Samples for Analysis

Samples for analysis were two spur gears from different model color inkjet printers. The specifications of these spur gears are shown in Table 1. The number of teeth on the spur gears is different. Spur marks from these spur gears were formed on pressure-sensitive film (Fujifilm, Prescale MS) by feeding the sheets of film into each printer. In this process, continuous spur mark lines of more than 10 spur gear rotations were sampled on the films.

The pitch data string of the spur mark lines was calculated from two-dimensional position data of the spur marks. The position of each spur mark was measured by a measurement microscope, which consisted of a two-axis micro-scanning stage (Chuo Precision Industrial, MSS-150) and a metaloscope (Nikon, Optiphoto 150). The micro-scanning stage was controlled to a resolution of 0.002 millimeters for each axis. Pitch data were calculated from the positional data of adjacent spur marks as shown in our previous study [23]. The length of a measured pitch data string was for 10 rotations of each spur gear.

Pitch data strings of limited length were prepared by extracting part of the pitch data string. In this study, eight pitch data lengths ranging from three to 10 times the number of spur gear teeth were prepared. Each pitch data string was extracted from the head of the pitch data string for 10 rotations of a spur gear.

Table 1. Specifications of spur gears

Spur gear	Manufacturer	Model	Number of teeth
A	Canon	BJF800	18
B	Seiko EPSON	PM-750C	24

In this study, the pitch data strings of length N are represented as follows:

$$\{x_1, x_2, \cdots, x_i, \cdots, x_N\}. \tag{1}$$

2.2 Maximum Entropy Spectral Analysis of Pitch Data Strings

The spectral components of each pitch data string were analyzed by the maximum entropy method (MEM) [24, 25]. MEM spectral analysis is considered to provide robust estimation even in the case of short data length. The maximum entropy spectrum at frequency f [cycle per tooth] was represented as follows,

$$P(f) = P_m \left/ \left|1 + \sum_{k=1}^{m} \gamma_{m,k} e^{-j \cdot 2\pi f k}\right|^2\right. . \tag{2}$$

Each parameter in Eq. (2) was calculated by Burg's algorithm [23,24] as follows,

$$
\left.
\begin{aligned}
P_m &= P_{m-1}\left(1 - \gamma_{m,m}^2\right), \\
\gamma_{m,m} &= -2 \sum_{i=1}^{N-m} b_{m,i} b'_{m,i} \left/ \sum_{i=1}^{N-m}\left(b_{m,i}^2 + b'^2_{m,i}\right)\right., \\
b_{m,i} &= b_{m-1,i} + \gamma_{m-1,m-1} \cdot b'_{m-1,i}, \\
b'_{m,i} &= b'_{m-1,i+1} + \gamma_{m-1,m-1} \cdot b_{m-1,i+1}, \\
b_{0,i} &= b'_{0,i} = x_i, \quad b_{1,i} = x_i, \quad b'_{1,i} = x_{i+1}.
\end{aligned}
\right\} \tag{3}
$$

The order "m" in Eq. (2) was determined by minimum AIC estimate (MAICE) [26]. AIC (Akaike's information criterion) [27] is widely used as a criterion for determining the number of parameters in statistical models, as it provides a reasonable model by balancing maximum likelihood against the number of parameters. In this study, AIC is represented as follows,

$$AIC = N \ln S_m^2 + 2m, \tag{4}$$

where,

$$S_m^2 = \sum_{i=m+1}^{N}\left(x_i + \gamma_{m,1} x_{i-1} + \gamma_{m,2} x_{i-2} + \cdots + \gamma_{m,m} x_{i-m}\right)^2 \left/ (N-m)\right. . \tag{5}$$

2.3 Estimation of Teeth Number by Maximum Entropy Spectrum

The number of spur gear teeth was estimated by the maximum entropy spectrum. The peak frequency of the spectrum was assumed to provide a frequency corresponding to the number of teeth. The peak was searched in the range of the frequency that corresponds to teeth numbers 15 to 48, based on transcendental information about spur gears. In this paper, we considered the peaks to be a signal to noise ratio (SNR) higher than 7, and also full width at half maximum (FWHM) as less than 0.02 cycles per tooth.

3 Results and Discussion

Figure 1 shows spur mark pitch data for 10 rotations of a spur gear. The pitch of each spur gear was not constant, but was different according to the section of spur marks. In the data for spur gear B, a periodic feature was easily observed because of the long pitch which appears at constant interval. However in the data for spur gear A, it was difficult to acquire periodicity at a glance.

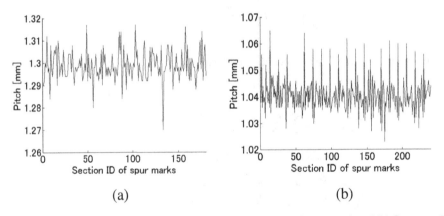

Fig. 1. Spur mark pitch data for 10 rotations of spur gear; (a) Spur gear A and (b) Spur gear B

Table 2 shows mean and standard deviation (S.D.) of the spur mark pitch data for 10 rotations of a spur gear. The mean of spur mark pitch was different between the two spur gears. The standard deviation was almost at the same level—smaller than 0.01.

Figure 2 shows the histogram of the spur mark pitch data for 10 rotations of spur gear. The distribution of spur mark pitch was different between the two spur gears. In spur gear A, pitch data showed an almost Gaussian distribution around a mean value except for extremely short pitch data observed occasionally. However in spur gear B, a minor group of long pitch data around 1.06 millimeters was observed in addition to a major group of pitch data around 1.04 millimeters. The minor group of longer pitch in spur gear B was considered to be noticed as a periodic feature in Fig. 1.

Table 2. Mean and standard deviation (S.D.) of spur mark pitch data for 10 rotations of a spur gear

Spur gear	Mean [mm]	S.D. [mm]
A	1.298	0.006
B	1.040	0.007

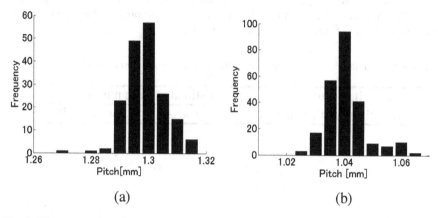

(a)   (b)

Fig. 2. Histogram of spur mark pitch data for 10 rotations of a spur gear; (a) Spur gear A and (b) Spur gear B

Table 3 lists the order in MEM determined by MAICE for each length of spur mark pitch data. In a case of pitch data longer than five times the number of teeth, the order was higher than the number of spur gear teeth. On the contrary, the order was lower than the number of spur gear teeth in a case where pitch data length was three times the number of teeth for spur gear A. It was also in the case pitch data length shorter than four multiple of teeth number for spur gear B.

Table 3. The order in MEM determined by MAICE for each length of spur mark pitch data

Length of pitch data [Multiple of teeth number]	Spur gear A	Spur gear B
3	16	16
4	18	16
5	20	35
6	20	25
7	24	25
8	19	25
9	19	25
10	19	48

Table 4 is the result of the estimation of the number of teeth of each spur gear for each length of spur mark pitch data. The estimation was successful within a deviation of one tooth for pitch data longer than five times the number of teeth. These results show that a successful estimation was achieved by using a shorter pitch data length than in previous study [22]. The required spur mark line length for the estimation was approximately 120 mm in this study. This is adequately included within A5 documents, and also it is shorter than the width of many banknotes and security documents. In addition, MEM spectral analysis is advantageous even in the case of short data length. Therefore, the result of this study indicates the proposed method can potentially be applied to forensic document examinations.

Table 4. Teeth number of spur gear estimated for each length of spur mark pitch data

Length of pitch data [Multiple of teeth number]	Estimated teeth number	
	Spur gear A	Spur gear B
3	-	-
4	-	-
5	19	24
6	18	24
7	18	25
8	18	25
9	18	24
10	18	25

However, there were failures in the estimation for spur mark pitch data that were shorter than four times the number of teeth. In these cases, a spectral peak was not obtained because of the absence of a spectral peak. Even in the case where a spectral peak exists, it is difficult to determine the peak for its weak and broad profile. The primary factor behind the failure was considered to be an inappropriate determination of the order in MEM. As shown in Table 3, the order determined by MAICE was smaller than the teeth number of each spur gear in these cases. Therefore, the nature of pitch data strings for which the current value has a strong relation with the previous value for the same number as teeth was not adequately considered in these cases.

In future work, appropriate determination of the order in MEM needs to be studied. In order to estimate the correct teeth number, the order needs to be higher than the teeth number of the spur gear. Although MAICE is a reasonable strategy for determining the order in MEM based on maximum likelihood and the number of parameters, it was not always appropriate for the objective of this study—especially in the case of short pitch data string. In this sense, better results may be provided by setting the order to be the highest value of teeth number among various spur gears.

Another problem is the estimation from intermittent spur mark pitch data. In actual forensic document analysis, spur mark lines left on inkjet-printed evidence are not always continuous, but sometimes occur intermittently. In order to apply the proposed method to a wide range of actual forensic document evidence, a technique for interpolating intermittent spur mark pitch data, and a method to estimate the periodicity originated in the spur mark teeth number needs to be developed in future work.

4 Conclusion

The effects of data length on an estimate of the number of teeth on spur gears of inkjet printers were studied in this paper. The correct number of teeth for each spur gear was estimated from spur mark pitch data longer than five times of the number of teeth by MEM spectral analysis. The result of this study achieved a successful estimation from shorter spur mark pitch data than in our previous study. Therefore, the proposed method was considered to improve the accuracy of printer model identification based on SCM in the field of forensic document analysis.

However, an estimation was not obtained from spur mark pitch data shorter than four times the number of teeth because of a failure in the MEM order determination by MAICE. In order to apply the proposed method to a wide range of actual forensic document evidence, a technique for interpolating intermittent spur mark pitch data, and a method for estimating the periodicity of the spur mark teeth number needs to be developed in the future.

References

1. Khanna, N., Mikkilineni, A.K., Chiu, G.T.C., Allebach, J.P., Delp, E.J.: Survey of Scanner and Printer Forensics at Purdue University. In: Srihari, S.N., Franke, K. (eds.) IWCF 2008. LNCS, vol. 5158, pp. 22–34. Springer, Heidelberg (2008)
2. Beusekom, J., Shafait, F., Breuel, M.: Document Signature Using Intrinsic Features for Counterfeit Detection. In: Srihari, S.N., Franke, K. (eds.) IWCF 2008. LNCS, vol. 5158, pp. 47–57. Springer, Heidelberg (2008)
3. Schulze, C., Schreyer, M., Stahl, A., Breuel, T.M.: Evaluation of Gray level-Features for Printing Technique Classification in High-Throughput Document Management Systems. In: Srihari, S.N., Franke, K. (eds.) IWCF 2008. LNCS, vol. 5158, pp. 35–46. Springer, Heidelberg (2008)
4. Pan, A.I.: Advances in Thermal Inkjet Printing. In: Proceedings of the SPIE, vol. 3422, pp. 38–44 (1998)
5. Doherty, P.: Classification of Ink Jet Printers and Inks. J. Am. Soc. Quest. Doc. Exam. 1, 88–106 (1998)
6. Lewis, J.A., Kondrat, M.: Comparative Examination of Black Ink Jet Printing Inks. In: Proceedings of the 49th Annual Meeting of American Academy of Forensic Sciences, NewYork, February 17-22, p. 182 (1997)
7. Mazzella, W.D.: Diode Array Micro Spectrometry of Colour Ink-jet Printers. J. Am. Soc. Quest. Doc. Exam. 2, 65–73 (1999)
8. Davis, J.E.: An Introduction to Tool Marks, Firearms and the Striagraph, pp. 3–6. Bannerstone House, Springfield (1958)
9. Totty, R.N., Baxendale, D.: Defect Marks and the Identification of Photocopying Machines. J. Forensic Sci. 21, 23–30 (1981)
10. James, E.L.: The Classification of Office Copy Machines from Physical Characteristics. J. Forensic Sci. 32, 1293–1304 (1987)

11. Gerhart, F.J.: Identification of Photocopiers from Fusing Roller Defects. J. Forensic Sci. 37, 130–139 (1992)
12. Hardcastle, R.A.: Progressive Damage to Plastic Printwheel Typing Elements. Forensic Sci. Int. 30, 267–274 (1986)
13. Moryan, D.D.: Cause of Typewriter Printwheel Damage Observed in the Questioned Document. J. Am. Soc. Quest. Doc. Exam. 1, 117–120 (1998)
14. Brown, J.L., Licht, G.: Using the ESDA to Visualize Typewriter Indented Markings. J. Am. Soc. Quest. Doc. Exam. 1, 113–116 (1998)
15. Mason, J.J., Grose, W.P.: The Individuality of Toolmarks Produced by a Label Marker Used to Write Extortion Notes. J. Forensic Sci. 32, 137–147 (1987)
16. LaPorte, G.M.: The Use of an Electrostatic Detection Device to Identify Individual and Class Characteristics on Document Produced by Printers and Copiers—a Preliminary Study. J. Forensic Sci. 49, 1–11 (2004)
17. Akao, Y., Kobayashi, K., Seki, Y.: Examination of spur marks found on inkjet-printed documents. J. Forensic Sci. 50, 915–923 (2005)
18. Akao, Y., Kobayashi, K., Sugawara, S., Seki, Y.: Discrimination of inkjet printed counterfeits by spur marks and feature extraction by spatial frequency analysis. In: van Renesse, R.F. (ed.) Optical Security and Counterfeit Deterrence Techniques IV, Proceedings of the SPIE, vol. 4677, pp. 129–137 (2002)
19. Akao, Y., Kobayashi, K., Seki, Y., Takasawa, N.: Examination of inkjet printed counterfeits by spur marks. In: Abstract of 4th Annual Meeting of Japanese Association of Science and Technology for Identification, Tokyo, p. 115 (1998) (in Japanese)
20. Maruyama, M.: Inventor. Seiko Epson Corp., assignee. Paper-pressing Mechanism for Ink Jet Printer. Japan patent 1814458, April 12 (1983)
21. Furukawa, T., Nemoto, N., Kawamoto, H.: Detection of Spur Marks Using Infrared Ray Scanner and Estimation for Out-of-Focus PSF. In: Abstract of 14th Annual Meeting of Japanese Association of Forensic Science and Technology, Tokyo, p. 190 (2008) (in Japanese)
22. Akao, Y., Kobayashi, K., Sugawara, S., Seki, Y.: Estimation of the number of spur teeth by determination of AR model order. In: Abstract of 6th Annual Meeting of Japanese Association of Science and Technology for Identification, Tokyo, p. 170 (2000) (in Japanese)
23. Akao, Y., Kobayashi, K., Sugawara, S., Seki, Y.: Measurement of pitch and mutual distance of spur marks by two-axes automatic scanning stage. In: Abstract of 5th Annual Meeting of Japanese Association of Science and Technology for Identification, Tokyo, p. 137 (1999) (in Japanese)
24. Burg, J.P.: Maximum entropy spectral analysis. In: 37th Annual International Meeting, Soc. of Explor. Geophys, Oklahoma City, Okla, October 31, (1967)
25. Burg, J.P.: Maximum Entropy Spectral Analysis. Ph.D dissertation, Stanford Univ. (1975)
26. Akaike, H.: A new look on the statistical model identification. IEEE Trans. Automat. Contr. AC-19, 716–723 (1974)
27. Akaike, H.: Information theory and an extension of the maximum likelihood principle. In: Petrov, B.N., Csaki, F. (eds.) 2nd International Symposium on Information Theory, Akademiai Kiado, Budapest, pp. 267–281 (1973)

Detecting the Spur Marks of Ink-Jet Printed Documents Using a Multiband Scanner in NIR Mode and Image Restoration

Takeshi Furukawa

Forensic Science Laboratory, Ibaraki Prefectural Police Headquarters,
978-6 Kasahara, Mito, Ibaraki, Japan
keikaso@pref.ibaraki.lg.jp

Abstract. Ink-jet printers are frequently used in crime such as counterfeiting bank notes, driving licenses, and identification cards. Police investigators required us to identify makers or brands of ink-jet printers from counterfeits. In such demands, classifying ink-jet printers according to spur marks which were made by spur gears located in front of print heads for paper feed has been addressed by document examiners. However, spur marks are significantly faint so that it is difficult to detect them. In this study, we propose the new method for detecting spur marks using a multiband scanner in near infrared (NIR) mode and estimations of point spread function (PSF). As estimating PSF we used cepstrum which is inverse Fourier transform of logarithm spectrum. The proposed method provided the clear image of the spur marks.

1 Introduction

Ink-jet printers have been widely used at home or in offices due to their low cost, their size and high quality of output. In the field of forensic document analysis, many counterfeit bank notes, forgery printing has been printed using ink-jet printers. Identifying makers or brands of ink-jet printers from counterfeits were required by police investigators. At first stage of the studies, chemical or spectral properties of ink-jets have been mainly focus on [1, 2, 3].

The method for comparing spur marks, illustrated in Fig.1, made by ink-jet printers was developed by Akao as an index of classifying ink-jet printers [4]. This method has been widely used in Japanese crime investigation laboratories. In the conventional method, spur marks are detected using a microscope, however, this method needs experts who train the techniques of observing spur marks.

We had reported the method that easily detected spur marks in ideal examples using high resolution flat bed image scanners [5]. When this method was used in actual ink-jet printed counterfeit documents it was difficult to distinguish between spur marks and dots printed by the ink-jet printers.

We proposed the new method that discriminates spur marks and dots using optical characteristics of ink-jet printer inks to NIR. Ink-jet printer inks, especially color dye inks are well used, are normally transmitted by NIR so that color dots printed by ink-jet printers are eliminated and only spur marks are visualized. NIR are readily illuminated

Z.J.M.H. Geradts, K.Y. Franke, and C.J. Veenman (Eds.): IWCF 2009, LNCS 5718, pp. 33–42, 2009.

from incandescent light bulbs. Recently, instead of the bulbs, light emission diodes (LED)s are used as light sources of NIR. The advantages of using LEDs are to avoid heat from light sources such as incandescent light bulbs and sizes of LEDs are compact. Our laboratory has a flat bed type image scanner with NIR and ultraviolet (UV) as a light source. This scanner has six kinds of light source, NIR (long, middle, short), UV, visible reflected light, and visible transmitted light. Accordingly the scanner was called multiband scanner. In detecting spur marks made by ink-jet printers, we can use the scanner in NIR mode. The multi band scanner in NIR mode is convenient to scan a wide area of objects, e.g., a piece of letter size paper compared to CCD or CMOS cameras with an attachment lens. The multi band scanner is based on a normal flat bed scanner with a visible light source. A light source that emitted NIR and UV was added to the scanner. A normal flat bed scanner has fixed focus for visible light as scanning reflective originals.

As shown in Fig.2, the multiband scanner has fixed focus structure for visible light and visible fluorescence to UV so that the focus is shifted from front plane to back plane when NIR is used because wave length of NIR is longer than that of visible light and the difference of the wave lengths mainly resulted in blurred images.

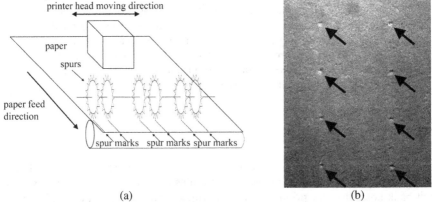

(a)   (b)

Fig. 1. (a) is the diagram of mechanism which makes spur marks. (b) is a typical example of lines of spur marks observed using microscope under oblique light illumination. Arrow in Fig.1 indicates spur marks. The magnification is 20 x.

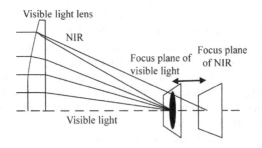

Fig. 2. The diagram of out-of-focus caused by the difference of the wave length between visible light and NIR

To put an attachment on the stage of the multi band scanner to lengthen focal distance for NIR improves the blurred image. However, the defect is that the lid of the multi band scanner is not shut so that visible light is not cut off.

Except for the difference of wave lengths a scanning mechanical system is also one of the causes of blurred. The plural causes are mixed so that the PSF of the scanner is complicated.

Instead of the above physical method, a computational simulation method is tried to this problem. This defect is improved using estimating the PSF that is borne by the multiband scanner in NIR mode. We propose the method which uses the multiband scanner in NIR mode and estimation of the PSF.

When the PSF is estimated, to find the nulls in a logarithm spectrum is important. Gennery estimated the PSF of out of focus from a radius of a dark ring in a logarithm spectrum of an observed image illustrated in Fig.3 [6]. The method proposed by Gennery is simple however it is difficult to measure a radius of a dark ring because the logarithm amplitude spectrum is not clear caused by noise.

In this study, we apply a new method for estimating a PSF using cepstrm. In next section we explain the detail of the method which is composed of the experimentation device and computational simulating principles.

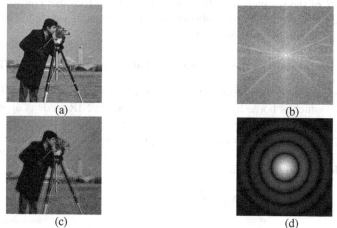

Fig. 3. Logarithm spectrum of images. (a) Original Image, (b) Logarithm spectrum of original image, (c) Defocus image, (d) Logarithm spectrum of defocus image.

2 Materials and Method

2.1 Samples for Analysis

We printed the test image, shown in Fig.6, using the ink jet printer (MP980, Canon). The papers we used were glossy papers (PMA3NSP1, Epson) and print mode was photo paper and print quality was fine. The image named Portrait was selected from Graphic technology-Prepress digital data exchange Standard colour image data (SCID), ISO/JIS-SCID because portraits printed in bank notes or facial photographs printed in driver licenses are frequently counterfeited. Fig.6 shows the image we used.

Fig.6 (a) is a TIFF format image that has a resolution of 400dpi and a pixel size of 2048 x 2560. This image file was printed out by the ink-jet printer mentioned above.

2.2 Visualization of Spur Marks by Multiband Scanner in NIR Mode

Image Fig.6 (b) was scanned by the multiband scanner (FM-10, Glory Ltd. Japan) in NIR mode. The multiband scanner, which equipped with NIR as light sources, was developed by Shimoyama et al. who work for forensic science laboratory, Hyogo prefectural Police Headquarters and Glory Ltd [7]. This scanner has multiband wave length light sources, i.e., visible light emitted by incandescent light bulbs, three kinds of wave length of NIR emitted by LEDs arrays, and UV emitted by fluorescent light tubes. The principle wave length of each light source is not described in the instruction manual [8] because the developers of the multiband scanner dose not desire to open the scanner specification and counterfeiters could use the same wave length for their forgeries when the detail of the scanner is public. We estimated the wave length of NIR that was emitted from the multiband scanner in NIR mode to compare images by using the video spectral comparator (VSC 6000, foster + freeman Ltd, UK). The result of the principle wave length estimated for suitable as detecting spur marks using mutiband scanner in NIR mode is approximate 900 nm that is most long wave length of NIR of the scanner. A well-known material of LEDs that emitted the radiation of principle wave length over 900nm is gallium arsenide (GaAs). The principle emission wave length of GaAS is 940nm.

2.3 Image Restoration by the Estimation of Point Spread Function (PSF)

Next, we try to restore the blurred NIR image using the estimation of the PSF. Estimating a PSF which is borne by the multi band scanner in NIR mode is applied to the blurred image. Next section we attempt to detect the spur marks printed by an ink-jet printer using the cepstrum. The blurred image g(x, y) is then equal to the convolution of the object intensity function f(x, y) and the PSF h(x, y) of the blurred system. All noise modeled as additive. We have

$$g(x,y) = \sum_{x',y'} f(x',y')h(x-x',y-y') + n(x,y). \tag{1}$$

In Fourier domain, this convolution is expressed as

$$G(u,v) = F(u,v)H(u,v) + N(u,v). \tag{2}$$

In this paper, the PSF is restricted to be a defocused lens system of the multi band scanner. As indicating Eq. (3) the PSF of the defocused lens system with a circular aperture can be approximated by cylinder whose radius r depends on the extent of the defocus.

$$h(x,y) = \begin{cases} \frac{1}{\pi r^2}; & x^2 + y^2 \le r^2 \\ 0; & x^2 + y^2 > r^2. \end{cases} \tag{3}$$

However, the multiband scanner has actually a complicate PSF that is combined not only the defocused lens system but also with a scanning mechanical system. In this preliminary study we try to resolve only the effect of the defocus lens system. When the PSF is transformed to Fourier domain, we obtain the optical transfer function (OTF) as H and indicating Eq, (4), (5).

$$H(u,v) = \frac{J_1(2\pi r f)}{\pi r f}.$$

(4)

$$f = \sqrt{(u^2 + v^2)}.$$

(5)

where r denotes the radius of circle of confusion, f denotes the radius of dark ring, J_1 is Bessel function of the first kind for integer order 1. This method Gennery proposed was applied to detect the blurred image of the spur marks by us using the multiband scanner in NIR mode [9]. In previous study it is difficult to estimate the radius of the dark ling because of the added noise. Accordingly we calculate the cepstrum of the blurred spur marks image. Cannon et, al estimated a PSF using a peak of a cepstrum, which was inverse Fourier transform of logarithm spectrum based on the idea of Gennery [10], [11]. This method provided appropriate results regardless of all simple procedures. The cepstrum is defined as Eq.6, where F^{-1} denotes the inverse Fourier integral transform.

$$C_g(p,q) = F^{-1}\{\log|G(u,v)|\}.$$

(6)

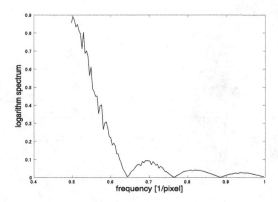

Fig. 4. The cross section of logarithm spectrum in out-of-focus blurred image

Fig.4 shows the logarithm spectrum of Fig.3 (c), (d). The horizontal axis shows frequency (1/pixel) and the vertical axis shows logarithm spectrum. The nulls found in profiles indicate the radius of dark ring. Fig.5 shows the cepstrum of Fig.4. The horizontal axis shows quefrency (pixel) and vertical axis shows cepstrum. The local minimum peak is found in the profile. The peaks indicate the radius of dark ring. We apply the inverse filter shown in Eq.7 to the blurred image.

$$H^{-1} = W(u,v) = \frac{\overline{H}(u,v)}{|H(u,v)|^2 + \beta^2} \tag{7}$$

The value β was 0.025 used, because the maximum of signal - noise ratio of CCDs is approximate 40. We try to detect spur marks by means of the method mentioned in previous section 2.2.

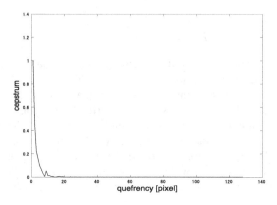

Fig. 5. The cross section of cepstrum in out-of-focus blurred image

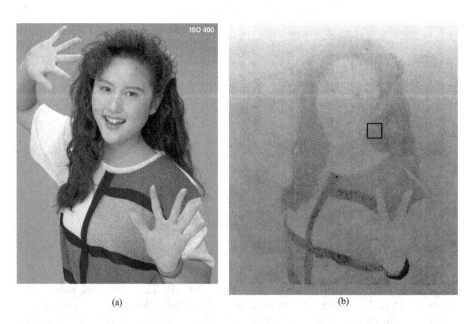

Fig. 6. Images used in the experiment. (a) Portrait, ISO, Original image is color image represented by CYMK. (b) The image printed using the ink-jet printer and the printed image was scanned by the multiband scanner in NIR mode. Fig 6. (b) shows that color inks are transmitted. The square in Fig (b) indicates the area which was restored.

3 Results and Discussion

Fig.6 shows the image we used. The square indicated in Fig.6 (b) is the area in which the spur marks found. The pixel size of this area was 256 x 256. The area magnified is shown in the Fig.7 (a). The image scanned by visible light corresponding to the image scanned by NIR is shown in Fig.10 (a). The subtle white dots are found in the Fig.10 (a), meanwhile the spur marks image represented as shades are found in Fig.7 (a) which was scanned using the multiband scanner in NIR mode. This image is blurred because of mainly out of focus depends on the differences of wave length between visible light and NIR.

Fig.7 (b) shows the logarithm spectrum of the blurred NIR image of Fig.7 (a). The dark rings in Fig.7 (b) are obscured by noise and it is difficult to measure the radius of dark rings.

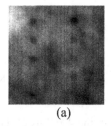

(a)   (b)

Fig. 7. (a) Image cut area indicated in Fig.6 (b) scanned by the multiband scanner in NIR mode, (b) Logarithm spectrum of Fig.7 (a)

Fig.8 shows the cross section of logarithm spectrum in Fig.7 (b). The horizontal axis shows frequency (1/pixel) and the vertical axis shows logarithm spectrum. The nulls in spectrum are not found because the most of the spectrum level is over 0.5.

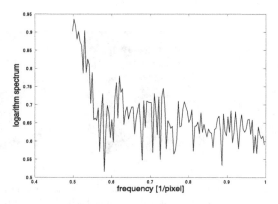

Fig. 8. The cross section of logarithm spectrum in image scanned by the multiband scanner in NIR mode

Fig.9 shows the cepstrum of cross section of the blurred image scanned by the multiband scanner in NIR mode. The horizontal axis shows quefrency (pixel) and the vertical axis shows cepstrum. As indicated in Fig.9, the subtle two local maximum peaks are observed at approximate 5 to 10 dots along horizontal axis. This extent of the values is used when we estimate the PSF.

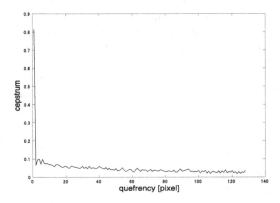

Fig. 9. The cross section of cepstrum in image scanned by the multiband scanner in NIR mode

Fig.10 indicates the image when the inverse filter with the 4 radius of dark ring is applied to the blurred image. Fig.10 (a) is the image scanned by visible light, Fig.10 (b) is the image scanned by the multiband scanner in NIR mode, and Fig.10 (c) is the image restored by the inverse filter. As shifting the left to the right rows, the images of the spur marks are gradually clear.

This study is the first step for detecting spur marks using multiband scanner in NIR mode. We will try to resolve a lot of problem i.e., applying to other kinds of papers, inks, and makers or brands of printers. The LEDs that emitted a most suitable wave

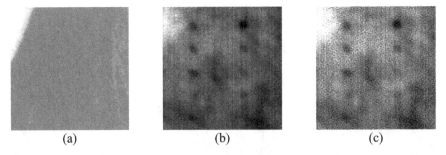

(a)   (b)   (c)

Fig. 10. Result of detecting the spur marks using the multiband scanner in NIR mode and image restoration. (a) is scanned with visible light, (b) is scanned with NIR, and (c) is scanned with NIR and restoration using cepstrum. Each image is 256 x 256 pixels.

length and a band width of NIR will have to be precisely investigated. The PSF that was had by the multiband scanner in NIR mode is complicate because not only the defocus lens system was caused but also a scanning mechanical system was caused. In this preliminary study the restoration image is not enough for measuring the pitch of spur marks and the distance between two lines of spur marks. In the feature work clearer image will be obtained when lots of these problems will be resolved. This study is the combination of the conventional methods however it is the first step for development of automatic detection and measurement system for spur marks.

4 Summary

We try to detect the spur marks using the multi band scanner in NIR mode. The illumination we used is NIR because ink-jet printer inks, especially color dye inks, are transmitted by NIR so that dots printed by ink-jet printers are eliminated and the spur marks observed as shades are obviously detected. In this process, we face the problem that the NIR images are blurred because of mainly out-of-focus depends on the differences between the length of visible light and that of the NIR. We apply the estimation of the PSF to this problem. It is difficult to estimate the PSF from the logarithm amplitude spectrum because of noise. We attempt to estimate the PSF using the cepstrum. This method is useful for detecting the peak which indicated the radius of dark ring in the logarithm spectrum calculated from the blurred spur marks image. The PSF is obtained from the radius of dark rings and we restore the spur marks image using the PSF.

References

1. Doherty, P.: Classification of Ink Jet Printers and Inks. J. Am. Soc. Quest. Doc. Exam. 1, 88–106 (1988)
2. Lewis, J.A.: Comparative Examination of Black Ink Jet Printing Inks. In: Proceedings of the 49th Annual Meeting of American Academy of Forensic Science, New York, February 17-22 (1972)
3. Mazzella, W.D.: Diode Array Micro Spectrometry of Colour Ink-jet Printers. J. Am. Soc. Quest. Doc. Exam. 2, 65–73 (1999)
4. Akao, Y., Kobayashi, K., Seki, Y.: Examination of Spur Marks Found on Inkjet-Printed Documents. Journal of Forensic Science 50(4), 915–923 (2005)
5. Nemoto, N.: Detection of Spur Marks Using a High Resolution Scanner. In: Abstract of 13th Annual Meeting of Japanese Association of Forensic Science and Technology, vol. 12, p. 182 (2007) (in Japanese)
6. Gennery, D.B.: Determination of Optical Transfer Function by Inspection of Frequency-Domain Plot. Journal of the Optical Society of America 63(12), 1571–1577 (1973)
7. Shimoyama, M., Kobayashi, K.: Development of Multiband Scanner for Examination Questioned Documents. Journal of Printing Science and Technology 45(4), 270–274 (2008) (in Japanese)
8. Multiband scanner FM-10 Instruction manual, p. 16. GLORY Ltd (2006) (in Japanese)

9. Furukawa, T., Nemoto, N., Kawamoto, H.: Detection of Spur Marks Using Infrared Ray Scanner and Estimation for Out-of-Focus PSF. In: Abstract of 14th Annual Meeting of Japanese Association of Forensic Science and Technology, vol. 13, p. 190 (2008) (in Japanese)
10. Cannon, M.: Blind Deconvolution of Spatially Invariant Image with Phase. IEEE Transaction on Acoustics, Speech, and Signal Processing ASSP-24(1), 58–63 (1978)
11. Shiqian, W., Zhongkang, L., Ee, P.O.: Blind Image Blur Identification in Cepstrum Domain. In: Computer Communication and Networks (ICCCN 2007), Proceeding of 16th International Conference, pp. 1166–1171 (2007)

A Computational Discriminability Analysis on Twin Fingerprints

Yu Liu and Sargur N. Srihari

Department of Computer Science and Engineering,
University at Buffalo, The State University of New York, Buffalo NY, USA
{yl73,srihari}@buffalo.edu

Abstract. Sharing similar genetic traits makes the investigation of twins an important study in forensics and biometrics. Fingerprints are one of the most commonly found types of forensic evidence. The similarity between twins' prints is critical establish to the reliability of fingerprint identification. We present a quantitative analysis of the discriminability of twin fingerprints on a new data set (227 pairs of identical twins and fraternal twins) recently collected from a twin population using both level 1 and level 2 features. Although the patterns of minutiae among twins are more similar than in the general population, the similarity of fingerprints of twins is significantly different from that between genuine prints of the same finger. Twins fingerprints are discriminable with a 1.5% ~ 1.7% higher EER than non-twins. And identical twins can be distinguished by examine fingerprint with a slightly higher error rate than fraternal twins.

1   Introduction

The studies of twins ramify into genetic [2], physiological [3], biochemical [1] and sociological aspects. There are two types of twins, dizygotic twins (commonly known as fraternal twins) and monozygotic twins, frequently referred to as identical twins. Fraternal twins occur when two fertilized eggs are implanted in the uterine wall at the same time. Identical twins develop when a single fertilized egg splits in two embryos. Because sharing a single zygote, identical twin individuals will have the same genetic makeup. Although there are lot of traits about twins that are not entirely genetic characteristics, twins still show high similarity in appearance, behavior, traits such as fingerprints, speech patterns and handwriting. Genetic and environmental similarities of twins allow studies such as the effectiveness of drugs, presence of psychological traits, etc. By examining the degree to which twins are differentiated, a study may determine the extent to which a particular trait is influenced by genetics or by the environment.

In forensics, questions concerned are how do the inherited and acquired traits of twins differ from those of singletons? How do the traits of identical twins differ from those of non-identical twins? By what means can twins be distinguished from other singletons? While having identical DNA makes it difficult for forensic scientists to distinguish between DNA from blood samples of identical twins. Few

Z.J.M.H. Geradts, K.Y. Franke, and C.J. Veenman (Eds.): IWCF 2009, LNCS 5718, pp. 43–54, 2009.
© Springer-Verlag Berlin Heidelberg 2009

twin studies have been carried out in any modality due to the lack of sufficient data. Such studies are important since any modality needs to be evaluated in conditions under which the possibility of error is maximum, i.e., the worst-case scenario. Satisfactory performance with twins strengthens the reliability of the method. It also establishes the degree of individuality of the particular trait. Such an individuality measure is relevant from the viewpoint of Daubert challenges in forensic testimony [6].

A forensic phonetic investigation on the speech patterns of identical and non-identical twins has indicated that they could be discriminated using Bayesian Likelihood Ratios [5]. A significant number of twin pairs (206) have been studied for handwriting [8]. These samples were processed with features extracted and conclusions drawn by comparing verification performances with twins and non-twins. In that study the conclusion was that twins are discriminable but less so than an arbitrary pair of individuals. An examination on palmprints generated from the same genetic information was carried out [7] and an automatic palmprint identification algorithm is provided to distinguish identical twins' palmprints. The study showed a significant correlation within identical twin matching (prints generated from the same genetic information).

The question for fingerprint is whether there exists a higher degree of similarity between individuals who are twins rather than when the individuals are not twins. The goal is to determine whether the fingerprints of twins are more similar to each other than in the case of the general population. A study on fingerprint of identical twins has been previously reported with a small data set of 94 pairs of index fingers [10]. Using a state-of-the-art fingerprint verification system it was concluded that identical twins are discriminable with slightly lower accuracy than non-twins. The marginal degradation in performance may be attributed to the dependence of minutiae distribution on fingerprint class. An earlier study [11] made use of fingerprints of 196 pairs of twins. In that study 196 comparisons of level 1 classification were made and when there was a match, a ridge count comparison was made. Level 2 (minutiae) comparison included only 107 pairs corresponding to identical twin fingerprints. In our previous twins' fingerprints study [12], we had compared the distribution of twins fingerprint classes to general population. The inferences that twins are different from genuines were made by statistic test of the similarity of matching score distribution of twins and that of the genuine distribution using a twins' database collected in 2003. In this study, we quantitively analyze the similarity of identical twins and fraternal twins using both level 1 and level 2 features on a later Twins' data set collected in 2007.

2   Twins Data Set

The data set used in this study consisted of friction ridge patterns of over five hundred pairs of twins. This data set was collected by the International Association for Identification (IAI) at a twins festival held in Twinsburg, Ohio in August 2007. The friction ridge images of 477 individuals including 227 sets of twins(188 sets of identical twins and 39 sets of fraternal twins) and one sets of triplets. For each individual there are ten fingerprints, thus making available 2,270 pairs of twin's fingers. In addition there are palm prints, DNA samples and latent prints collected, but not used in this study. In our earlier study [12], similar twins' data set used are collected by IAI in 2003. The electronic capture of fingerprints were scanned by Smith Heiman scanner and processed with Cogent System, Inc. Software. images were obtained at 500 dpi (with image size of 800 × 750). The 2007 Twins data set is of a higher quality than the 2003 Twins data set.

A meta-data table accompanying each folder of fingerprint images gives the demographic information for the individual, code for the individual and a pointer to his/her twin. The demographic information consists of age, gender, hair color, racial characteristics, whether twins are identical or fraternal, and handedness (left/right). The distribution of ages of the twins shows that the twins in the study are predominantly in their adolescent years. Thus the quality of their fingerprints can be expected to be good.

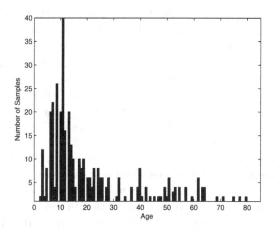

Fig. 1. Distribution of twins' ages in database: a predominance of younger ages is observed

3 Experiment Results

Friction ridge patterns contained in fingerprints can be analyzed at several levels of features. Level 1 features correspond to visually observable characteristics commonly used in fingerprint classification, i.e., arch, loop, whorl are the most common fingerprint ridge flow types. Level 2 features correspond to minutiae, which are primarily points corresponding to ridge endings and ridge bifurcations. Level 3 features include pores within ridges, ridge widths, and shapes. The analysis reported here was done using level 1 and level 2 features.

3.1   Level 1 Study

The first study was to determine the similarities at Level 1 or fingerprint classes. A number of studies have shown that there are correlation in the fingerprint class (level 1 features) of identical twin fingers [9]. Correlation based on other generic attributes of the fingerprint such as ridge count, ridge width, ridge separation, and ridge depth has also been found to be significant in identical twins. In our study, the twins fingerprint images are labeled according to one of six categories: arch, tented arch, right loop, left loop, whorl and twin loop [13]. The overall distribution of the six level 1 features provides an indication of how frequently each class is encountered in the Twins database, i.e., arch (5%), tented arch (3%), left loop (32%), right loop (34%), whorl (20%) and twin loop (6%).

The analysis consisted of determining as to how often the prints of the same finger in a pair of twins matched and a comparison with the case of non-twins. For twins, we count the total number of finger pairs that belong to the same fingerprint class. Among the 227 pairs of twins, the percentage of times twins had the same class label for a given finger was 62.78%. If identical twins is considered, the percentage of same level 1 was 66.71% as against 46.41% for fraternal twins. For non-twins, if we compute the probability that any two fingerprints share the same class label in the Twin' data set is as $p_{arch}^2 + p_{tented\ arch}^2 + p_{left\ loop}^2 + p_{right\ loop}^2 + p_{whorl}^2 + p_{twin\ loop}^2$, that is 0.265. Then the probability of any two person who are not twins having the same level 1 type is no more than 26.5%, is much lower than the probability of twins' match. In [10], the probability that two randomly picked fingerprints that have the same class label is computed to be 27.18% according to five major fingerprint classes in the index finger based on an unpublished FBI report using database of $22,000,000$ human-classified fingerprints. Thus we can conclude that twins are more than twice as likely as non-twins in matching level 1 features. Figure 2 shows an examples of exact same class labels of 10 pairs of prints belonged to one identical twins.

Our results demonstrate a significant correlation in level 1 features of twins, especially identical twins. However, level 1 features are used only as a coarse method of eliminating candidates from a large database as in automatic fingerprint identification systems (AFIS). It has no implication on the discriminability of twins since level 1 features are not used in fingerprint identification.

3.2   Level 2 Study

The most important part of the study from the point of view of identification involves level 2 features since they are what are used by AFIS systems in fingerprint matching. Level 2 features consist of minutiae which are either ridge endings or ridge bifurcations. Each minutia is represented by a 3-tuple (x, y, θ) representing its position and orientation in the fingerprint image. The question to be examined is as to whether the fingerprints match when minutiae are used as features. To quantitatively measure similarity, we use an AFIS type algorithm that extracts minutiae and obtains a score from the comparison. The MIN-DTCT algorithm for detecting minutiae and the Bozorth matcher [14], which provides

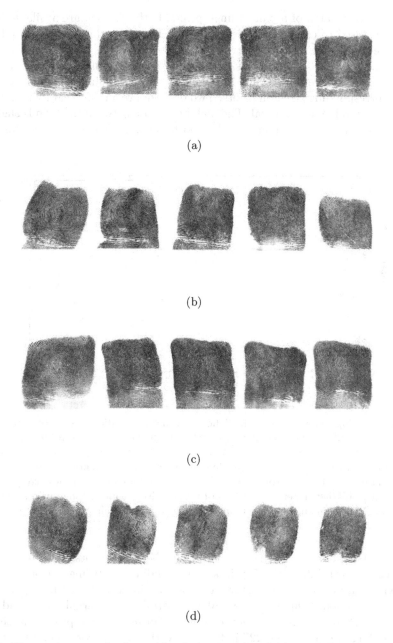

(a)

(b)

(c)

(d)

Fig. 2. Fingerprint examples for an identical twin. (a) and (c) are 10 fingerprint images of an individual A, (b) and (d) are 10 fingerprint images of the corresponding twin B. The twins have exactly identical fingerprint classes for both hands. (Class labels are: (left-hand) whorl, twin loop, right loop, whorl, right loop; (right-hand) twin loop, whorl, left loop, left loop, left loop)

a score for the match of a pair of fingerprints, both of which are available from NIST, was used to compare fingerprint pairs. The Bozorth score is typically in the range of 0-50 for impostor scores and can be as high as 350 for genuine (see Figure 3). Other AFIS algorithms have similar scores but have different ranges. Figure 4 shows three fingerprints images and extracted minutiae superimposed on the corresponding skeleton prints. Two of them are an identical twin and the other is an unrelated individual. The matching score between the twin is slightly higher than that between non-twin, but both impostor scores are less than 50.

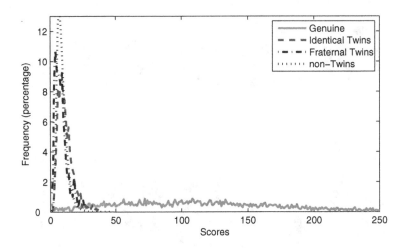

Fig. 3. Matching score distributions.(The following distributions were obtained: non-Twin: An individual's fingerprint was matched with the corresponding fingerprint of all other people who were not his/her twin. Identical: The fingerprint of an individual was matched with the corresponding fingerprint of his/her identical twin(twin-twin match). Fraternal: The fingerprint of an individual was matched with the corresponding fingerprint of his/her fraternal twin(twin-twin match). Genuine: Fingerprints Pairs in FVC2002 DB1 belonging to a same finger were compared against each other.)

All 227 pairs of twins was used to carry out the experiments. The fingerprints are rolled fingerprints with 10 prints(corresponding to 10 fingers) per person. For twin-twin match, the impostor score distribution is obtained from the total 2,270 comparisons. Out of these 780 were prints of fraternal twins and the remaining 1,880 were those of identical twins. For non-twin impostor match, we compare fingerprints among 227 individuals without their corresponding twins. An individual's fingerprints are matched with the corresponding fingerprints of all other people who is unrelated to him/her. The total number of comparisons was $256,510$ $((10 \times 227 \times 226)/2)$. In order to get a score distribution of genuine match (pairs of fingerprints that belong to the same finger were compared against each other), we used FVC2002 DB1 data set, due to the lack of multiple rolled fingerprint samples of the same finger in the Twins' data set. Similar to the

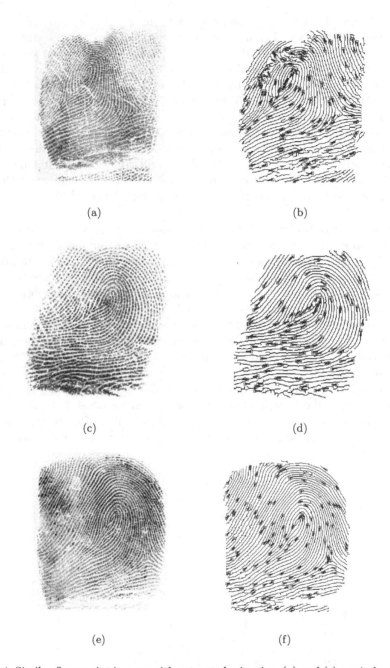

(a)

(b)

(c)

(d)

(e)

(f)

Fig. 4. Similar fingerprint images with extracted minutiae. (a) and (c) are index fingerprint images of an identical twin, (e) is an index fingerprint image of another unrelated person. (b), (d), (f) are the corresponding minutiae extracted (with their locations and orientations indicated). The matching score between (a) and (c) is 47; while the matching score between (a) and (e) is 7.

Twins' data set, the data set contains live scan images at resolution of 500 dpi (with image size of 388 × 374) of forefinger and middle finger of both the hands of each individual. A total of 100 fingers were presented with 8 samples of each finger constituting a total of 800 prints. The genuine scores are obtained by comparing each fingerprint with the rest of the 7 impressions resulting a 2800 (100×(8×7)/2 matchings. Figure 3 shows these distributions. A slight shift in the twins impostor distribution in comparison to the non-twins impostor distribution can be observed, which gives an impression that distinguishing twins are slightly harder than distinguishing any two random people by matching their fingerprints since false positive rate is increased. Besides, the distribution of matching score of identical twins lie a little bit right to the that of fraternal twins indicating a slightly increase of score on average level for matching identical twin.

The Kolmogorov-Smirnov(KS) Goodness-Of-Fit Test. By statistically comparing probability of similarity among the score distributions for twins, non-twin and genuine matching, we need to answer a question: "Can we disprove, with a certain required level of significance, the null hypothesis that the two distributions are drawn from the same population?"[15]. The test first obtains the cumulative distribution functions of each of the distributions to be compared, and then computes the statistic, D, as the maximum value of the absolute difference between two of the cumulative distribution functions to be compared. What makes the KS statistic useful is that the distribution in the case of the null hypothesis (data sets drawn from the same distribution) can be calculated, at least useful approximation, thus giving the significance of any observed nonzero value of D. For example, if comparing twins with non-twin, the KS statistic D is given by $D = \max_{-\infty<x<\infty} |S_{twins}(x) - S_{non-twin}(x)|$ (see Figure 5), where x denotes the matching score and $S_{twins}(x)$ and $S_{non-twin}(x)$ are cumulative distribution

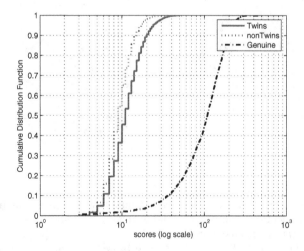

Fig. 5. Cumulative distribution function for Kolmogorov-Smirnov test

functions for twins and non-twin. The significance level of an observed value of D is given approximately [15].

We computed the significance of D between any two distributions shown in Figure 5. The resulting p-value of the indicates a significance level that the null hypothesis is accepted. For both twin-genuine and nontwin-genuine comparisons, the computed probability of any observed nonzero value of D is 0. For twin-nontwin comparison, the p-value is slightly larger than 0 (5.9957e-082) which indicates a 99.99% confidence level of rejecting the null hypothesis. The conclusion can be drawn as there is a significantly different between matching scores distribution of twins and that of non-twins. This distribution difference can also be observed from Figure 5. Moreover, the p-value of comparing identical twins and fraternal twins is found to be 0.012. Although it is not strong enough to reject null hypothesis if the 1% significance level is considered. The result still indicates that identical twins share more similarity in the level 2 feature than the fraternal twins. As a whole, the similarity of fingerprints of twins is different from that between genuine prints of the same finger.

Receiver Operating Characteristics(ROC) Analysis. In a biometric verification system, the system decision is normally regulated by a threshold t. The scores provided by the matcher can be thresholded to provide a hard decision of being the same or different.

The ROC curves are in Figure 6 for matching twins and non-twins, as well as the system decision thresholds shown along the curve. The ROC curve of the matching twin pairs (including identical twins and fraternal twins) are lower than that of non-twin pairs. There is a trade off between FAR and FRR that both depend on the choice of threshold. The equal error rate (EER) with

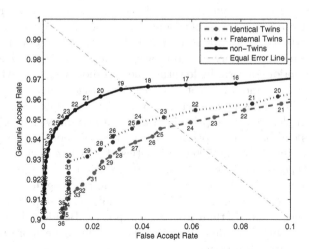

Fig. 6. ROC curves of twins and non-twin matching. Operating points(thresholds) are also shown along the curves.

Table 1. False Acceptance Rate, False Rejection Rate and Average Error Rate with different thresholds (19~25) for twin and non-twin matching in the Twins data set (the Equal Error Rate is underlined)

Threshold	FRR (%)	non-twin		fraternal		identical	
		FAR (%)	AER (%)	FAR (%)	AER (%)	FAR (%)	AER (%)
19	3.5	3.15	<u>3.33</u>	11.03	7.26	13.62	8.56
20	3.86	2.33	3.09	9.49	6.67	11.12	7.49
21	4.21	1.75	2.98	8.46	6.34	9.63	6.92
22	4.54	1.29	2.91	6.15	5.34	8.14	6.34
23	4.89	0.96	2.93	4.87	<u>4.88</u>	6.91	5.90
24	5.14	0.72	2.93	3.85	4.49	5.96	5.55
25	5.46	0.52	2.99	3.59	4.53	4.73	<u>5.09</u>

the corresponding operation threshold point for all three cases is shown in Table 1. The average error rate for matching identical twins is roughly 1.7% higher than non-twin matching and the average matching error for fraternal twins is 1.5% higher than non-twin matching if the operating point is selected based on EER.

Because the matching score distribution for twins are slight closer to the genuine score distribution comparing to the non-twin matching scores, FAR is increased as a given threshold but the FRR remains unchanged (same threshold for different matching scenarios lie on a same horizontal line in Figure 6). Table 1 shows the FAR and FRR values with different thresholds ranging from 19 to 25 for non-twin and twin matching. The discriminability study of Twins data shows that for a given threshold at 25, at which a EER can be achieved and there is a chance of 5.46% that the same finger can be mistakenly rejected, a FAR of 0.52% for non-twin matching and a FAR of 4.73% for identical twins indicates fingerprints of twins are 9 times more likely than non-twin to be falsely accepted. While In cases of forensic applications such as criminal identification, where FRR is a major concern, by lowering the threshold to 19, a FAR of 3.15% for non-twin matching and a FAR of 13.62% for identical twins indicates finger-prints of twins are 4 times more likely to to be falsely accepted than unrelated people. For example, in a fingerprint identification system with 1 million non-twin individuals enrolled, 31500 people will be falsely accepted at the threshold of 19. If such system contains 500,000 identical twins fingerprint pairs, there would be 136000 people be falsely accepted. Figure 7 illustrates the ratio of false match (and false non-match) between twins and non-twins against the choice of system threshold. With the increase of the threshold, twin imposters are more likely to be falsely accepted than non-twin imposter. Besides, due to the ge-netically similarity between identical twins, they tend to be more hard to be distinguished than fraternal twins.

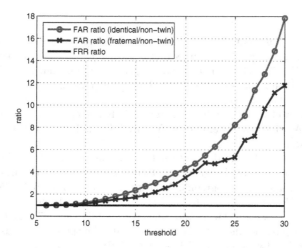

Fig. 7. Ratio between FAR and FRR for twins and non-twin matching

4 Summary and Conclusion

A study of the discriminability of the fingerprints of twins using a latest set of samples has been presented. Live scans and younger ages of the subjects ensured good quality prints thereby allowing the focus to be on the inherent individuality of fingerprints and one that was not affected by image quality issues. The similarities of the fingerprints of both identical twins and fraternal twins were studied using fingerprint features at levels 1 and 2. The level 1 results show that identical twins finger's have a significant correlation in fingerprint class leading to a higher probability of having the same type of ridge flow(62.78% for all twins) than in the case of non-twins (26.5%). Level 2 features were studied using a minutiae-based matching algorithm which provides a similarity score. Distributions of scores were compared using the Kolmogorov-Smirnov test. The system errors at different operating points are analyzed for the cases of twin-twin match and non-twin matching. The statistical inferences from the level 2 study are: comparing to fingerprint imposters from random unrelated people, it is more difficult to distinguish fingerprints of twins. The EER for matching twins is roughly 1.5% ∼ 1.7% higher than non-twin matching. However, the similarity of fingerprints of twins is significantly different from that between genuine prints of the same finger. Besides, due to the inherit similarity between identical twins, they can be distinguished by examine fingerprint with a slightly higher error rate than fraternal twins. This leads to a conclusion differing from the previous 2003 Twins data set. Similar to the study on 2003 Twins data set, the implications are that there is more similarity between twin fingers than in the case of two arbitrary fingers, and that twins can be successfully discriminated using fingerprints.

References

1. Evans, A., van Baal, G.C.M., McCarron, P., de Lange, M., Soerensen, T.I.A., de Geus, E.J.C., Kyvik, K., Pedersen, N.L., Spector, T.D., Andrew, T., Patterson, C., Whitfield, J.B., Zhu, G., Martin, N.G., Kaprio, J., Boomsma, D.I.: The Genetics of Coronary Heart Disease: The Contribution of Twin Studies. Twin Research 2003 6(5), 432–441 (2003)
2. Rubocki, R., McCue, B., Duffy, K., Shepard, K., Shepherd, S., Wisecarver, J.: Natural DNA Mixtures Generated in Fraternal Twins in Utero. J. For. Sci. 46(1), 120–125 (2001)
3. Spitz, E., Mountier, R., Reed, T., Busnel, M.C., Marchaland, C., Robertoux, P.L., Carlier, M.: Comparative Diagnoses of Twin Zygosity by SSLP Variant Analysis, Questionnaire, and Dermatoglyphic Analysis. Behav. Genet. 26(1), 55–63 (1996)
4. Goldberg, S., Perrotta, M., Minde, K., Corter, K.: Maternal Behaviour and Attachment in Low Birthweight Twins and Singletons. Child Dev. 57(1), 34–46 (1986)
5. Loakes, D.: A forensic phonetic investigation into the speech patterns of identical and non-identical twins. IJSLL 15.1, 97–100 (2008)
6. Daubert v Merrell Dow Pharmaceuticals, Inc. 509 U.S. 579, 113 S.Ct. 2786 (1993)
7. Kong, A.W.-k., Zhang, D., Lu, G.: A study of identical twins' palmprints for personal verification. Pattern Recognition 39, 2149–2156 (2006)
8. Rihari, S.N., Srinivasan, C.H.: On the Discriminability of the Handwriting of Twins. J. For. Sci. (Accepted for publication)
9. Jain, A.K., Prabhakar, S., Pankanti, S.: Twin Test: On Discriminability of Fingerprints. In: Bigun, J., Smeraldi, F. (eds.) AVBPA 2001. LNCS, vol. 2091, pp. 211–216. Springer, Heidelberg (2001)
10. Jain, A.K., Prabhakar, S., Pankanti, S.: On similarity of identical twin fingerprints. Pattern Recognition 35, 2653–2663 (2002)
11. Lin, C.H., Liu, J.H., Osterburg, J.W., Nicol, J.D.: Fingerprint Comparison 1: Similarity of Fingerprints. J. For. Sci. 1982 27(2), 290–304 (1982)
12. Srihari, S.N., Srinivasan, H., Fang, G.: Discriminability of Fingerprints of Twins. Journal of Forensic Identification 58, 109–129 (2008)
13. Maltoni, D., Maio, D., Jain, A.K., Prabhakar, S.: Handbook of Fingerprint Recognition, p. 176. Springer, New York (2003)
14. Garris, M.D., Watson, C.I., McCabe, R.M., Wilson, C.L.: Users Guide to NIST Fingerprint Image Software (NFIS); NISTIR 6813; NIST. In: U.S. Dept. of Commerce: Gaithersburg, MD, p. 20
15. Press, W.H., Flannery, B.P., Teukolsky, S.A., Vetterling, W.T.: Numerical Recipes in C: The Art of Scientific Computing, 2nd edn. Cambridge University Press, Cambridge (1992)

Probability of Random Correspondence for Fingerprints

Chang Su and Sargur N. Srihari

Computer Science and Engineering Department
University at Buffalo, Buffalo NY 14228, USA
{changsu,srihari}@cedar.buffalo.edu

Abstract. Individuality of fingerprints can be quantified by computing the probabilistic metrics for measuring the degree of fingerprint individuality. In this paper, we present a novel individuality evaluation approach to estimate the probability of random correspondence (PRC). Three generative models are developed respectively to represent the distribution of fingerprint features: ridge flow, minutiae and minutiae together with ridge points. A mathematical model that computes the PRCs are derived based on the generative models. Three metrics are discussed in this paper: (i) PRC of two samples, (ii) PRC among a random set of n samples (nPRC) and (iii) PRC between a specific sample among n others (specific nPRC). Experimental results show that the theoretical estimates of fingerprint individuality using our model consistently follow the empirical values based on the NIST4 database.

1 Introduction

We consider the evaluation of the relevant probabilities in a commonly used forensic modality – that of fingerprints. Their use in human identification has been based on two premises, that, (i) they do not change with time and (ii) they are unique for each individual. While in the past, identification based on fingerprints had been accepted by courts, more recently their use has been questioned under the basis that the premises stated above have not been objectively tested and error rates not been scientifically established [1]. Though the first premise has been accepted, the second one on individuality is being challenged.

Fingerprint individuality studies date back to the late 1800s. More than twenty models have been proposed to establish the improbability of two random people (or fingers) have the same fingerprint [2]. These models can be classified into five different categories: grid-based [3,4], ridge-based [4], fixed probability [5], relative measurement [6] and generative [7,8,9]. All models try to quantify the uniqueness property, e.g., the probability of false correspondence.

Features for representing fingerprints are classified into three types [10]. Level 1 features provide class-characterization of fingerprints based on ridge flow. Fig.1 shows five primary classes: whorl, left loop, right loop, arch and tent. Level 2 features, which are more useful for identification, are also known as minutiae. The minutiae correspond to ridge endings and ridge bifurcations. A minutia is

Z.J.M.H. Geradts, K.Y. Franke, and C.J. Veenman (Eds.): IWCF 2009, LNCS 5718, pp. 55–66, 2009.
© Springer-Verlag Berlin Heidelberg 2009

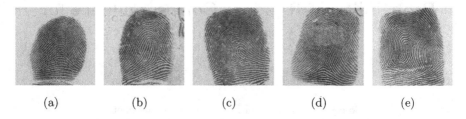

(a)   (b)   (c)   (d)   (e)

Fig. 1. Examples of five main types of ridge flow, referred to as Level 1 features: (a) arch, (b) left loop, (c) right loop. (d) tented arch, and (e) whorl

represented by its location and direction; direction is determined by the ridge ending at the location (Figure 2(a)). Automatic fingerprint matching algorithms use minutiae as the salient features [11], since they are stable and are reliably extracted. For some cases in which minutiae can not provide sufficient information, ridge points are also taken into account. The ridge points are shown in Figure 2(b). Level 3 features, such as pores and scars are ancillary features. In this paper, only level 1 and level 2 features are considered. Our goal is to model the distribution of fingerprint features and then estimate the individuality of fingerprints based on the models.

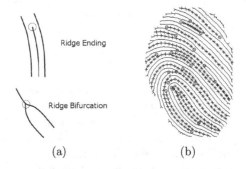

(a)   (b)

Fig. 2. Representation of fingerprints using minutiae and ridgepoints: (a) locations of ridge endings and ridge bifurcations are indicated by circles, and (b) ridge points are indicated by stars in a skeletonized fingerprint image

This paper is the subsequence of our work on generative models on fingeprints [9]. We measure individuality by 3 metrics: PRC, nPRC and specific PRC on level 1 and level 2 features. A noval PRC caculate algorithm is proposed to get rid of the dependence on the fingerprint matchers. The rest of this paper is organized as follows: Section 2 describes the generative models proposed for the ridge flow types, minutiae only and a combination– representative ridges. Methods for estimating individuality and experimental results are described in Section 3. Summary is given in Section 4.

2   Generative Models for Fingerprints

2.1   Distribution of Ridge Flow Types

A simple distribution of the Level 1 ridge flow types is obtained by counting the relative frequency of each of the primary and secondary types in a fingerprint database. In one such evaluation [16] loops account for 64% of the fingers, with the secondary types being: 30% left loops, 27% right loops and 7% double loops. Arches account for 18% of the primary types, with the seondary types being: plain arches (13%) and tented arches (5%). Whorls account for the remainder of the Level 1 types (19%).

2.2   Distribution of Minutiae

Each minutia is represented as $\mathbf{x} = (s, \theta)$ where $s = (x_1, x_2)$ is its location and θ its direction. The distribution of minutiae location is shown in Fig. 3(a). The minutiae is taken from 2000 fingerprints in NIST4 database.

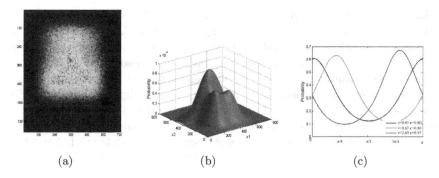

(a)   (b)   (c)

Fig. 3. Model for minutiae distribution: (a) Distribution of minutia location from NIST special database, (b) Three-dimensional plot of mixture Gaussian model for minutia location with three components, (c) von Mises distributions of minutiae orientation for each of the three components, where the green curve corresponds to the upper cluster, blue the lower left cluster and red the lower right cluster

Since minutia location has a multimodal distribution as shown in Fig.3(a), a mixture of K Gaussians is a natural approach. Furthermore since minutiae orientation is a periodic variable, it is modeled by a *circular normal* or von Mises distribution[12,13]. Such a model is better than mixtures of hyper-geometric and binomial distributions [7,8]. We write the distribution of minutiae as

$$p(\mathbf{x}|\Theta) = \sum_{k=1}^{K} \pi_k \cdot \mathcal{N}(s|\mu_k, \Sigma_k) \cdot \mathcal{V}(\theta|\nu_k, \kappa_k), \qquad (1)$$

where K is the number of mixture components, π_k are non-negative component weights that sum to one, $\mathcal{N}(s|\mu_k, \Sigma_k)$ is the bivariate Gaussian probability density function of minutia with mean μ_k and covariance matrix Σ_k, $\mathcal{V}(\theta|\nu_k, \kappa_k)$ is the von Mises probability density function of minutiae orientation with mean angle ν_k and precision κ_k, and $\Theta = \{\pi_k, \mu_k, \Sigma_k, \nu_k, \kappa_k, \rho_k\}$ where $k = 1, 2, .., K$ is the set of all parameters of the k Gaussian and von Mises distributions.

Rather than using the standard form of the von Mises distribution for the range $[0, 2\pi]$, since minutiae orientations are represented as being in the range $[0, \pi)$, we use the alternate form [13] as follows

$$\mathcal{V}(\theta|\nu_k, \kappa_k, \rho_k) = \rho_k v(\theta) \cdot I\{0 \leq \theta < \pi\} + (1 - \rho_k)v(\theta - \pi) \cdot I\{\pi \leq \theta < 2\pi\} \quad (2)$$

where $I\{A\}$ is the indicator function of the condition A,

$$v(\theta) \equiv v(\theta|\nu_k, \kappa_k) = \frac{1}{\pi I_0(\kappa_i)} exp[\kappa_i cos2(\theta - \nu_k)], \quad (3)$$

minutiae arising from the kth component have directions that are either θ or $\theta + \pi$ and the probabilities associated with these two occurrences are ρ_k and $1 - \rho_k$ respectively.

Since fingerprint ridges flow smoothly with very slow direction changes, direction of neighboring minutiae are strongly correlated, i.e., minutiae that are spatially close tend to have similar directions with each other. However, minutiae in different regions of a fingerprint tend to be associated with different region-specific minutiae directions thereby demonstrating independence [14,15]. The model allows ridge orientations to be different at different regions (different regions can be denoted by different components) while it makes sure that nearby minutiae have similar orientations (as nearby minutiae will belong to the same component). The parameters for generative models are learned from NIST4 fingerprint database. We show the results in Figure 3(b) 3(c).

2.3   Distribution of Representative Ridges

In the model just discussed, only minutiae was used in the framework of generative models for fingerprints. Models based purely on minutiae may be sufficient to model biometric scenarios where finger-prints are obtained in controlled conditions [7][8], but are insufficient to model forensic scenarios where latent prints are lifted off of surfaces. In such cases, ridge details provide vital information about fingerprints. Verification systems using minutiae together with ridge information are more accurate than using minutiae alone. Also, any generative model that makes use of ridge details can only be a better representation of the generative model for fingerprints. With this motivation, the distribution of ridge information is embedded into generative models [9].

Since nearby ridges always maintain similar trend, too much redundancy would be involved if all the ridges are modeled. We only consider ridges where the minutiae lie. These ridges are called *representative ridges*. A representative ridge contains two parts, the minutiae and the following ridge. The latter part

Fig. 4. Representation of ridge points in polar coordinates

is represented as a set of ridge points sampled at equal interval of inter ridge width [18]. The length of the ridge l_r is defined as the number of ridge points that could be sampled on the ridge. Three types of ridges are defined: (i) short: $l_r \leq L/3$, (ii) medium: $L/3 < l_r \leq 2L/3$ and (iii) long: $l_r > 2L/3$, where L is the average ridge length of the top 10% longest representative ridges in the fingerprints database, e.g. in NIST4, L is 18. By choosing the value of maximum ridge length L as the average of long ridges in the database, unusually long ridges caused by artifacts is avoided. The three possible ridge length types can be associated with any representative ridge. Without loss of generality, we can assume that there exist only three possible ridge length types corresponding to a representative ridge. For ridges with different lengths, different ridge points are picked as anchors. For medium ridges, $\lfloor L/3 \rfloor^{th}$ ridge point is picked and for long ridges, both $\lfloor L/3 \rfloor^{th}$ and $\lfloor 2L/3 \rfloor^{th}$ are picked. No ridge point is chosen for short ridges. The rationale for choosing these two ridge points is described in [18].

Let $\mathbf{x} = \{\mathbf{x}_m, \mathbf{x}_r\}$ denote the feature vector of a representative ridge, where minutiae feature vector $\mathbf{x}_m$ is given by $\{s_m, \theta_m\}$ and ridge feature vector $\mathbf{x}_r$ is represented by a ridge point set given by $\{\mathbf{x}_{ri} : i \in A \land i \leq l_r\}$ where $\mathbf{x}_{ri}$ is the feature vector of the ith ridge point, A is the anchor point index set $\{\lfloor L/3 \rfloor, \lfloor 2L/3 \rfloor\}$ and l_r is the length of the ridge. In contrast to the model for minutiae, ridge points are represented in polar coordinates as shown in Figure 4. The location of ridge point s_{ri} is given by $\{r_i, \phi_i\}$ and the direction of ridge point is θ_i. Thus the ridge point $\mathbf{x}_{ri}$ is represented as the combination of location and direction $\{r_i, \phi_i, \theta_i\}$ where r_i is the distance from the ith ridge point to the minutia, ϕ_i is the positive angle required to reach the ith ridge point from the polar axis.

The generative model is based on the distribution of representative ridges. Mixture distributions consisting of K_i components, $i = 1, 2, 3$, is used to model representative ridges of three ridge length types. Each component is distributed according the density of the minutiae and density of ridge points. Assuming that the minutiae and ridge points are independent, representative ridge distribution is given by

$$p(\mathbf{x}|\Theta) = \begin{cases} p(l_r) \cdot \sum_{g=1}^{K_1} \pi_k p_k(s_m, \theta_m | \Theta_k) & l_r \leq \frac{L}{3} \\[2ex] p(l_r) \cdot \sum_{k=1}^{K_2} \pi_k p_k(s_m, \theta_m | \Theta_k) \\ \cdot p_k(r_{\lfloor \frac{L}{3} \rfloor}, \phi_{\lfloor \frac{L}{3} \rfloor}, \theta_{\lfloor \frac{L}{3} \rfloor} | \Theta_k) & \frac{L}{3} < l_r < \frac{2L}{3} \\[2ex] p(l_r) \cdot \sum_{k=1}^{K_3} \pi_k p_k(s_m, \theta_m | \Theta_k) \cdot p_k(r_{\lfloor \frac{L}{3} \rfloor}, \phi_{\lfloor \frac{L}{3} \rfloor}, \theta_{\lfloor \frac{L}{3} \rfloor} | \Theta_k) \\ \cdot p_k(r_{\lfloor \frac{2L}{3} \rfloor}, \phi_{\lfloor \frac{2L}{3} \rfloor}, \theta_{\lfloor \frac{2L}{3} \rfloor} | \Theta_k) & l_r \geq \frac{2L}{3} \end{cases}$$

$$(4)$$

The first condition corresponds to minutiae alone, the second to minutia and one ridge point, and the third to minutia and two ridge points. $p_k(s_m, \theta_m | \Theta_k)$ is the component distribution for minutiae location s_m and direction θ_m and $p_k(r_i, \phi_i, \theta_i | \Theta_k)$ is the component distribution for the ith ridge point. They are defined as in (5) and (6) respectively.

$$p_k(s_m, \theta_m | \Theta_k) = \mathcal{N}(s_m | \mu_{mk}, \Sigma_{mk}) \cdot \mathcal{V}(\theta_m | \nu_{mk}, \kappa_{mk}, \rho_{mk}) \tag{5}$$

$$p_k(r_i, \phi_i, \theta_i | \Theta_k) = p_k(r_i, \phi_i | \mu_{ik}, \Sigma_{ik}, \nu_{ik}^{\phi}, \kappa_{ik}^{\phi}, \rho_{ik}^{\phi}) \cdot \mathcal{V}(\theta_i | \nu_{ik}^{\theta}, \kappa_{ik}^{\theta}, \rho_{ik}^{\theta}) \tag{6}$$

where $p_k(s_m, \theta_m | \Theta_k)$ can be caculated by Eq. 1 and $p_k(r_i, \phi_i, \theta_i | \Theta_k)$ is the product of the probabilities of ridge point locations and directions, where $\mathcal{V}(\theta_i | \nu_{ik}^{\theta}, \kappa_{ik}^{\theta}, \rho_{ik}^{\theta})$ presents the distribution of the ridge point direction, θ_i is the direction of the ith ridge point and $p_k(r_i, \phi_i | \mu_{ik}, \Sigma_{ik}, \nu_{ik}^{\phi}, \kappa_{ik}^{\phi}, \rho_{ik}^{\phi})$ is the distribution of ridge point location given by

$$p_k(r_i, \phi_i | \mu_{ik}, \sigma_{ik}, \nu_{ik}^{\phi}, \kappa_{ik}^{\phi}, \rho_{ik}^{\phi}) = \mathcal{N}(r_i | \mu_{ik}, \sigma_{ik}) \cdot \mathcal{V}(\phi_i | \nu_{ik}^{\phi}, \kappa_{ik}^{\phi}, \rho_{ik}^{\phi}) \tag{7}$$

where $\mathcal{N}(r_i | \mu_{ik}, \sigma_{ik})$ is a univariate Gaussian distribution whose mean μ_{ik} and variance σ_{ik} are learnt from a fingerprint database.

3   Evaluation of PRCs

As measures of individuality, three probabilities can be defined: PRC, nPRC and Specific nPRC. These definitions are further described below.

1. PRC: probability that two randomly chosen samples have the same measured value $\mathbf{x}$ within specified tolerance ϵ.
2. nPRC: the probability that among a set of n samples, some pair have the same value $\mathbf{x}$, within specified tolerance, where $n \geq 2$. Since there are $\binom{n}{2}$ pairs involved this probability is higher than PRC. Note that when $n = 2$, PRC=nPRC.
3. Specific nPRC: the probability that in a set of n samples, a specific one with value $\mathbf{x}$ coincides with another sample, within specified tolerance. Since we are trying to match a specific value $\mathbf{x}$, this probability depends on the probability of $\mathbf{x}$ and is generally smaller than PRC. The exact relationship with respect to PRC depends on the distribution of $\mathbf{x}$.

We note that the first two measures, PRC and nPRC characterize the forensic modality as described by a set of measurements and furthermore, the second is a function of the first. The third measure specific nPRC characterizes specific evidence, e.g., a specimen. It is dependent on the specific value as well as the distribution of the measurement.

3.1   Ridge Flow Types

Assuming that fingerprints are distinguished by 6 secondary types, the PRC for ridge flow types is calculated by definition of PRC. Given type frequencies mentioned in section 2.1, we have PRC value $p_\epsilon = 0.2233$. The nPRC and specific nPRC can be simply computed from PRC value. Table 1 and table 2 show the nPRC and Specific nPRC with different n.

Table 1. Ridge Flow Types: nPRC

n	2	3	4	5	6	7	8
nPRC	0.2233	0.5314	0.7805	0.9201	0.9774	1	1

Table 2. Ridge Flow Types: Specific nPRC belong to class LL = left loop, RL = right loop, DL = double loop, PA = plain arch, TA = tented arch and W = whorl

n	Specific nPRC					
	LL	RL	DL	PA	TA	W
10	0.9596	0.9411	0.4796	0.7145	0.3698	0.8499
20	0.9989	0.9975	0.7481	0.9291	0.6226	0.9818

Level 1 features are clearly broad class characteristics which are useful for exclusion of individual fingers but not by themselves useful for the tasks of verification, identification and individualization.

3.2   Minutiae Only

To compute the PRCs for minutiae, we first define correspondence, or match, between two minutiae. Let $\mathbf{x_a} = (s_a, \theta_a)$ and $\mathbf{x}_b = (s_b, \theta_b)$ be a pair of minutiae. The minutiae are said to correspond if for tolerance $\epsilon = [\epsilon_s, \epsilon_\theta]$,

$$|s_a - s_b| \leq \epsilon_s \text{ and } |\theta_a - \theta_b| \leq \epsilon_\theta \tag{8}$$

where $|s_a - s_b|$ is the Euclidean distance between the minutiae location $s_a = (x_{a1}, x_{a2})$ and $s_b = (x_{b1}, x_{b2})$.

Then, the probability that a random minutia $\mathbf{x}_a$ would match a random minutia $\mathbf{x}_b$ is given by

$$p_\epsilon(\mathbf{x}) = p(|\mathbf{x}_a - \mathbf{x}_b| \le \epsilon|\Theta)$$

$$= \int\limits_{\mathbf{x}_a} \int\limits_{|\mathbf{x}_a - \mathbf{x}_b| \le \epsilon} p(\mathbf{x}_a|\Theta)p(\mathbf{x}_b|\Theta)d\mathbf{x}_a d\mathbf{x}_b \qquad (9)$$

where Θ is the set of parameters describing the distribution of the minutiae location and direction.

Finally, the PRC, or the probability of matching at least $\hat{m}$ pairs of minutiae within ϵ between two randomly chosen fingerprint f_1 and f_2 is calculated as

$$p_\epsilon(\hat{m}, m_1, m_2) = \binom{m_1}{\hat{m}}\binom{m_2}{\hat{m}}\hat{m}! \cdot p_\epsilon(\mathbf{x})^{\hat{m}}(1 - p_\epsilon(\mathbf{x}))^{(m_1-\hat{m})\cdot(m_2-\hat{m})} \qquad (10)$$

where m_1 and m_2 are numbers of minutiae in fingerprints f_1 and f_2, $p_\epsilon(\mathbf{x})^{\hat{m}}$ is the probability of matching $\hat{m}$ specific pairs of minutiae between f_1 and f_2, $(1 - p_\epsilon(\mathbf{x}))^{(m_1-\hat{m})\cdot(m_2-\hat{m})}$ is the probability that none of minutiae pair would match between the rest of minutiae in f_1 and f_2 and $\binom{m_1}{\hat{m}}\binom{m_2}{\hat{m}}\hat{m}!$ is the number of different match sets that can be paired up.

Given n fingerprints and assuming that the number of minutiae in a fingerprint m can be modeled by the distribution $p(m)$, the general PRCs $p(n)$ is given by

$$p(n) = 1 - \bar{p}(n) = 1 - (1 - p_\epsilon)^{\frac{n(n-1)}{2}} \qquad (11)$$

where p_ϵ is the probability of matching two random fingerprint from n fingerprints. If we set the tolerance in terms of number of matching minutiae to $\hat{m}$, p_ϵ is calculated by

$$p_\epsilon = \sum_{m_1' \in M_1} \sum_{m_2' \in M_2} p(m_1')p(m_2')p_\epsilon(\hat{m}, m_1', m_2') \qquad (12)$$

where M_1 and M_2 contain all possible numbers of minutiae in one fingerprint among n fingerprints, and $p_\epsilon(\hat{m}, m_1', m_2')$ can be calculated by Eq. 10.

Given a specific fingerprint f, the specific nPRCs can be computed by

$$p(f, n) = 1 - (1 - p(f))^{n-1} \qquad (13)$$

where $p(f)$ is the probability that $\hat{m}$ pairs of minutiae are matched between the given fingerprint f and a randomly chosen fingerprint from n fingerprints.

$$p(f) = \sum_{m' \in M} p(m')\binom{m'}{\hat{m}} \cdot \sum_{i=1}^{\binom{m_f}{\hat{m}}} p(f_i)$$

$$= \sum_{m' \in M} p(m')\binom{m'}{\hat{m}} \cdot \sum_{i=1}^{\binom{m_f}{\hat{m}}} \prod_{j=1}^{\hat{m}} p(\mathbf{x}_{ij}|\Theta) \qquad (14)$$

where M contains all possible numbers of minutiae in one fingerprint among n fingerprints, $p(m')$ is the probability of a figerprint having m' minutiae in n

fingerprints, m_f is the number of minutiae in the given fingerprint f, minutiae set $f_i = (\mathbf{x}_{i1}, \mathbf{x}_{i2}, ..., \mathbf{x}_{i\hat{m}})$ is the subset of the minutiae set of given fingerprint and $p(f_i)$ is the joint probability of minutiae set f_i based on learned generative model.

3.3 Minutiae with Ridge Points

When ridge information is considered, a representative ridge is denoted by $\mathbf{x} = \{\mathbf{x}_m, \mathbf{x}_r\}$, where $\mathbf{x}_r = \{\mathbf{x}_r^i : i \in \{\lfloor L/3 \rfloor, \lfloor 2L/3 \rfloor\} \wedge i \leq l_i\}$. The representative ridge $\mathbf{x}_a$ matchs the representative ridge $\mathbf{x}_b$ with tolerance ϵ if

$$|\mathbf{x}_{ma} - \mathbf{x}_{mb}| \leq \epsilon_m \wedge |\mathbf{x}_{ra} - \mathbf{x}_{rb}| \leq \epsilon_r \tag{15}$$

where $|\mathbf{x}_{ma} - \mathbf{x}_{mb}| \leq \epsilon_m$ is define by Eq.8 and $|\mathbf{x}_{ra} - \mathbf{x}_{rb}| \leq \epsilon_r$ is defined as

$$|\mathbf{x}_{ra} - \mathbf{x}_{rb}| \leq \epsilon_r \equiv (\forall i \in A)|r_a^i - r_b^i| \leq \epsilon_r \wedge |\phi_a^i - \phi_b^i| \leq \epsilon_\phi \wedge |\theta_a^i - \theta_b^i| \leq \epsilon_\theta \tag{16}$$

where A is the anchor point index set and the tolerance can be grouped together as $\epsilon = \{\epsilon_s, \epsilon_\theta, \epsilon_r, \epsilon_\phi\}$.

Then, the probability that a random representative ridge $\mathbf{x}_a$ would match a random representative ridge $\mathbf{x}_b$ is given by

$$p_\epsilon(\mathbf{x}) = p(|\mathbf{x}_a - \mathbf{x}_b| \leq \epsilon|\Theta)$$
$$= \int_{\mathbf{x}_a} \int_{|\mathbf{x}_a - \mathbf{x}_b| \leq \epsilon} p(\mathbf{x}_a|\Theta) p(\mathbf{x}_b|\Theta) d\mathbf{x}_a d\mathbf{x}_b \tag{17}$$

where Θ is the set of parameters describing the distribution of the representative ridges.

The nPRC and specific nPRC with ridge information can be caculated by Eq.11, 13 and 14.

3.4 Experiments and Results

Parameters of the two fingerprint generative models introduced in Sections 2.2 and 2.3 were evaluated using the NIST fingerprint database. The NIST fingerprint database, NIST Special Database 4, contains 8-bit gray scale images of randomly selected fingerprints. Each print is 512×512 pixels with 32 rows of white space at the bottom of the print. The entire database contains fingerprints taken from 2000 different fingers with 2 impression of the same finger. Thus, there are a total of 2000 fingerprints using which the model has been developed. The number of components K for the mixture model was found after validation using k-means clustering.

Values of PRC p_ϵ are calculated using the formula introduced in Section 3. The tolerance is set at $\epsilon_s = \epsilon_r = 10$ pixels and $\epsilon_\theta = \epsilon_\phi = \pi/8$. For comparison, the empirical PRC $\hat{p}_\epsilon(\mathbf{x})$ was calculated with the same tolerance. To compute $\hat{p}_\epsilon(\mathbf{x})$,

Table 3. PRC for different fingerprint matches with varying m_1(number of minutiae/ridges in fingerprint f_1),m_2 (number of minutiae/ridges in fingerprint f_2) and $\hat{m}$ (number of matched minutiae/ridges)

			Minutiae Only		Minutiae and Ridge Points	
m_1	m_2	$\hat{m}$	p_ϵ	$\hat{p}_\epsilon$	p_ϵ	$\hat{p}_\epsilon$
16	16	12	2.2×10^{-5}	5.8×10^{-6}	2.0×10^{-10}	3.7×10^{-15}
36	36	12	2.1×10^{-3}	3.4×10^{-3}	3.0×10^{-4}	1.2×10^{-5}
56	56	12	6.4×10^{-3}	6.6×10^{-3}	1.3×10^{-3}	1.4×10^{-4}
76	76	12	9.1×10^{-3}	1.1×10^{-2}	2.5×10^{-3}	5.5×10^{-4}

the empirical probabilities of matching a minutiae pair or ridge pair between imposter fingerprints are calculated first by

$$\hat{p}_\epsilon(\mathbf{x}) = \frac{1}{I} \sum_{i=1}^{I} \frac{\hat{m}_i}{m_i \times m_i'} \tag{18}$$

where I is the number of the imposter fingerprints pairs, $\hat{m}_i$ is the number of matched minutiae or ridge pairs and m_i and m_i' are the numbers of minutiae or pairs in each of the two fingerprints. Then, the empirical PRC $\hat{p}_\epsilon$ can be calculated by Eq.10.

Both the theoretical and empirical PRCs are given in Table 3. The PRCs are calculated through varying numbers of minutiae or ridges in two randomly chosen fingerprint f_1 and f_2 and the number of matches between them. We can see that more minutiae or ridges the template and input fingerprint have, higher the PRC is.

Table 4. nPRCs with varying m and $\hat{m}$ given $n = 100,000$ fingerprints

No. of	No of	Minutiae only	Minutiae and Ridge Points
minutiae/ridges m	matches $\hat{m}$	$p(n)$	$p(n)$
46	46	5.0680×10^{-34}	6.7329×10^{-49}
	26	1.1842×10^{-3}	1.7579×10^{-11}
36	36	1.5551×10^{-22}	7.5000×10^{-35}
	16	1	3.4852×10^{-1}
16	16	1.3464×10^{-1}	4.460×10^{-8}
	6	1	1

Based on the PRC value, nPRC can be computed. Table 4 shows the nPRCs in 100,000 fingerprints through varying number of minutiae or ridges in each fingerprint m and number of matches $\hat{m}$.

The specific nPRCs are also computed by (13) and given by Table 5. Here three fingerprints are chosen as query prints and they are shown in Figure 5.

(a)   (b)   (c)

Fig. 5. Three specific fingerprints (from the same finger) used to calculate probabilities:
(a) good quality full print F_1, (b) low quality full print F_2, and (c) partial print F_3

Table 5. Specific nPRCs for fingerprints in Fig. 5 with $n = 100,000$ fingerprints

Fingerprint f	No. of m/r m_f	No. of m/r m	No. of matches $\hat{m}$	Minutiae only $p(n,f)$	Minutiae and Ridge Points $p(n,f)$
F_1	41	40	31	3.0327×10^{-78}	8.5202×10^{-190}
		20	12	1.2995×10^{-18}	2.6310×10^{-28}
F_2	26	20	12	1.1057×10^{-21}	5.9308×10^{-45}
		10	4	8.3675×10^{-2}	6.6823×10^{-4}
F_3	13	20	12	2.0588×10^{-24}	1.4185×10^{-76}
		10	4	4.1707×10^{-2}	1.2057×10^{-3}

The first one is a full print in good quality, the second one is a full print in low
quality and the third one is a partial print. The specific nPRCs are calculated
through varying number of minutiae/ridges in each template fingerprint (m)
and the number of matches $(\hat{m})$, assuming that the number of fingerprints in
template database (n) is $100,000$. The numbers of minutiae/ridges in 3 given
query fingerprint m_f are 41, 26 and 13.

4   Summary

Generative models of individuality attempt to model the distribution of features
and then use the models to determine the probability of random correspondence. We have proposed such models of fingerprint individuality for ridge flow,
minutiae and representative ridges. Individuality is evaluated in terms of three
probability measures: probability of random correspondence (PRC) between two
individuals, general probability of random correspondence (nPRC) between two
individuals among a group of n individuals and specific probability of random
correspondence (specific nPRC) which is the probability of matching a given individual among n individuals. We perform the individuality estimation on NIST4
dataset. As expected, we found that PRCs are reduced when more features are
incorporated in the model. The proposed ridge information model offers a more
reasonable and more accurate fingerprint representation. The results provide a

much stronger argument for the individuality of fingerprints in forensics than previous generative models.

References

1. United States Court of Appeals for the Third Circuit: USA v. Byron Mitchell. No. 02-2859 (2003)
2. Stoney, D.A.: Measurement of Fingerprint Individuality. In: Lee, H., Gaensslen, R.E. (eds.) Advances in Fingerprint Technology. CRC Press, Boca Raton (2001)
3. Galton, F.: Finger Prints. McMillan, London (1892)
4. Roxburgh, T.: Galton's work on the evidential value of fingerprints. Indian Journal of Statistics 1, 62 (1933)
5. Henry, E.R.: Classification and Uses of FingerPrints, p. 54. Routledge & Sons, London (1900)
6. Trauring, M.: Automatic comparison of finger-ridge patterns. Nature, 197 (1963)
7. Pankanti, S., Prabhakar, S., Jain, A.K.: On the individuality of fingerprints. IEEE Transactions on Pattern Analysis and Machine Intelligence 24(8) (2002)
8. Dass, S.C., Zhu, Y., Jain, A.K.: Statistical Models for Assessing the Individuality of Fingerprints. In: Fourth IEEE Workshop on Automatic Identification Advanced Technologies, pp. 3–9 (2005)
9. Su, C., Srihari, S.N.: Generative Models for Fingerprint Individuality using Ridge Models. In: Proceedings of International Conference on Pattern Recognition. IEEE Computer Society Press, Los Alamitos (2008)
10. Ashbaugh, D.R.: Quantitative-Qualitative Friction Ridge Analysis: An Introduction to Basic and Advanced Ridgeology. CRC Press, Boca Raton (1999)
11. Watson, C., Garris, M., Tabassi, E., Wilson, C., McCabe, R., Janet, S.: User's Guide to NIST Fingerprint Image Software, vol. 2, pp. 80–86. NIST (2004)
12. Bishop, C.: Pattern Recognition and Machine Learning. Springer, New York (2006)
13. Mardia, K.V., Jupp, P.E.: Directional Statistics. Wiley, Chichester (2000)
14. Chen, J., Moon, Y.-S.: A Minutiae-based Fingerprint Individuality Model. In: Computer Vision and Pattern Recognition, CVPR 2007 (2007)
15. Hsu, R.L.V., Martin, B.: An Analysis of Minutiae Neighborhood Probabilities. In: Biometrics: Theory, Applications and Systems (2008)
16. Srihari, S.N., Cha, S., Arora, H., Lee, S.J.: Discriminability of Fingerprints of Twins. Journal of Forensic Identification 58(1), 109–127 (2008)
17. Tabassi, E., Wilson, C.L., Watson, C.I.: Fingerprint Image Quality. In: NISTIR 7151, National Institute of Standards and Technology (August 2004)
18. Fang, G., Srihari, S.N., Srinivasan, H., Phatak, P.: Use of Ridge Points in Partial Fingerprint Matching. In: Biometric Technology for Human Identification IV, pp. 65390D1–65390D9. SPIE (2007)

A New Computational Methodology for the Construction of Forensic, Facial Composites

Christopher Solomon[1,2], Stuart Gibson[1], and Matthew Maylin[1]

[1] School of Physical Sciences
University of Kent
Canterbury, Kent CT2 7NJ, U.K.
[2] School of Physics
National University of Ireland, Galway
Galway, Republic of Ireland
c.j.solomon@kent.ac.uk, s.j.gibson@kent.ac.uk,
matthew.maylin@visionmetric.com

Abstract. A facial composite generated from an eyewitness's memory often constitutes the first and only means available for police forces to identify a criminal suspect. To date, commercial computerised systems for constructing facial composites have relied almost exclusively on a feature-based, 'cut-and-paste' method whose effectiveness has been fundamentally limited by both the witness's limited ability to recall and verbalise facial features and by the large dimensionality of the search space. We outline a radically new approach to composite generation which combines a parametric, statistical model of facial appearance with a computational search algorithm based on interactive, evolutionary principles. We describe the fundamental principles on which the new system has been constructed, outline recent innovations in the computational search procedure and also report on the real-world experience of UK police forces who have been using a commercial version of the system.

Keywords: Evolutionary algorithms, facial composites, EFIT-V.

1 Introduction

Facial composite systems are used by the police to construct facial likenesses of suspects based on eyewitness testimony. The desired outcome is that the generated composite image be of sufficient accuracy that subsequent display to members of the public will result in recognition followed by the apprehension of the suspect. From a scientific and technological perspective, there are critical aspects to consider in the design of an effective composite system. It should operate in a manner that is both sympathetic to the cognitive processes involved in human face processing and be capable of producing realistic and accurate composite images. However, the majority of facial composite systems currently employed by police services require the witness to view and select individual features, (such as overall face shape, eyes, nose, mouth etc) from a stored database and then to electronically overlay these to achieve the best

Z.J.M.H. Geradts, K.Y. Franke, and C.J. Veenman (Eds.): IWCF 2009, LNCS 5718, pp. 67–77, 2009.

likeness. Previous research has suggested that the need for the witness to recall and verbally describe the face to the operator could be the weakest link in the composite construction process [1].

The subject of this paper is the technical design, computational implementation and field-use of a facial composite system, developed for use in criminal investigations at the University of Kent. Unlike traditional feature-based methods, the approach described here exploits global (whole face) facial characteristics and allows a witness to produce plausible, photo-realistic face images in an intuitive way using an interactive evolutionary algorithm. The earliest attempt to apply evolutionary methods to composite construction originates from the work of Johnston [6]. A broadly similar approach to composite construction using evolutionary methods has also been described by Frowd et al [5]. The implementation presented in this paper is a hybrid system, exploiting both evolutionary and systematic search methods. It had its beginnings in preliminary work reported by Solomon et al [2] and by Gibson et al [3] and has now grown into a successful commercial system, EFIT-V [4].

In section 2 of this paper, we outline the basis of our computer model of facial appearance. This model is generative and we show how new, highly plausible examples of human faces can be produced through statistical sampling of the constructed model. The computational search algorithm to find the desired facial likeness is inspired by evolutionary methods and our basic methodology is presented in section 3. In section 4, we outline a recent innovation to the evolutionary search procedure which accelerates the convergence to the target face by systematically exploiting witness input as to what the face is *not like*. In the following section, an account is given of a number of deterministic tools for altering the facial appearance which prove to be of importance in practice and which can further accelerate the convergence to the target. Finally, in section 6, we offer a summary of the system, its performance in real-world use and implications for the future.

2   Generative, Computational Model of Facial Appearance

The objective of a statistical appearance model is to learn the full range of natural shape and texture variation that occurs within a given pattern class. 'Active' appearance models have previously been used for computer pattern recognition applications [7] in which a face is synthesized and morphed to match a target face. In such cases, an automated, iterative fitting algorithm is employed to minimize a well-defined metric. In facial composite applications however, the fitting procedure must be guided by the witness's response to face stimuli and thus is not amenable to precise expression in mathematical form. The essential steps in the construction of the model (see [7] or [8] for full mathematical details) are –

i)   The annotation of a (preferably large) training sample with landmarks to define the shape vectors of each example.

ii)   The extraction of corresponding pixel intensities from each example to form shape-normalised texture vectors.

iii)   Separate principal component analyses (PCA) on the shape and texture vectors to produce the dominant modes of variation and to allow a reduction in the dimensionality of the shape and texture representations.

iv)   A further PCA to decorrelate the reduced shape and texture vectors to form appearance vectors.

Accordingly, an appearance model (AM) was constructed using a sample of 2729 faces face images, both male and female and comprising a broad range of ages and different ethnic origins. Using the AM, a compact vector representation, $\mathbf{c} = \begin{pmatrix} c_1 & c_2 & \cdots & c_N \end{pmatrix}$ of the appearance of an out-of-sample face can be obtained. The relationship between these appearance parameters and the corresponding shape $\mathbf{x}$ and texture parameters $\mathbf{g}$ is expressed in matrix-vector form as follows –

$$\mathbf{c} = \left\{ \mathbf{Q}^{\mathrm{T}} \begin{bmatrix} w\mathbf{b}_s \\ \mathbf{b}_g \end{bmatrix} = \mathbf{Q}^{\mathrm{T}} \begin{bmatrix} w\mathbf{P}_s^{\mathrm{T}}(\mathbf{x} - \overline{\mathbf{x}}) \\ \mathbf{P}_g^{\mathrm{T}}(\mathbf{g} - \overline{\mathbf{g}}) \end{bmatrix} \right\} \tag{1}$$

$$\mathbf{x} = \overline{\mathbf{x}} + \mathbf{P}_s \mathbf{b}_s \tag{2}$$

$$\mathbf{g} = \overline{\mathbf{g}} + \mathbf{P}_g \mathbf{b}_g \tag{3}$$

In equations (1) – (3), the columns of $\mathbf{Q}, \mathbf{P}_s$ and $\mathbf{P}_g$ correspond to the appearance, shape and texture principal components of the sample data respectively, $\overline{\mathbf{x}}$ and $\overline{\mathbf{g}}$ are the average shape and texture vectors, $\mathbf{b}_s$ and $\mathbf{b}_g$ are vectors of shape and texture parameters respectively and w is a scalar chosen to equalise the variance over the shape and texture samples. Given a set of appearance parameters $\mathbf{c} = \begin{pmatrix} c_1 & c_2 & \cdots & c_N \end{pmatrix}$, equation (1) can be inverted for $\mathbf{b}_s$ and $\mathbf{b}_g$ and substitution into equations (2) and (3) produces an actual facial image as defined by the shape and texture vectors $\{\mathbf{x}, \mathbf{g}\}$. In anticipation of our evolutionary search algorithm described in section 3, we will henceforth refer to the appearance vector $\mathbf{c}$ as the *genotype* and to the corresponding $\{\mathbf{x}, \mathbf{g}\}$ produced by equations (1) – (3) as the *phenotype*.

New, plausible examples of faces may be produced by sampling for the elements of the genotype $\mathbf{c}$ from the learned, multivariate distribution of the sample. In practice, this is conveniently and accurately approximated by an independently distributed, multivariate normal density function. Sampling for faces associated with a chosen demographic group can thus be generated by using an appropriate sub-sample model embedded in the parameter space of the standard form -

$$N(\mathbf{c}; 0, \mathbf{\Sigma}) = (2\pi)^{-\frac{n}{2}} |\mathbf{\Sigma}|^{-\frac{1}{2}} \exp\left\{ -\frac{1}{2}(\mathbf{c} - \overline{\mathbf{c}})^{\mathrm{T}} \mathbf{\Sigma}^{-1}(\mathbf{c} - \overline{\mathbf{c}}) \right\} \tag{4}$$

where $\mathbf{\Sigma} = \left\langle (\mathbf{c} - \overline{\mathbf{c}})(\mathbf{c} - \overline{\mathbf{c}})^{\mathrm{T}} \right\rangle_{sample}$ is the sample covariance matrix of the appearance parameters averaged over the sub-sample and $\overline{\mathbf{c}}$ is the sub-sample average.

In practice, equation 4 is sampled by generating N standard, independently distributed normal variables and then applying the appropriate linear transform. Examples of faces synthesized from different demographic sub-sample models using this method are presented in figure 1.

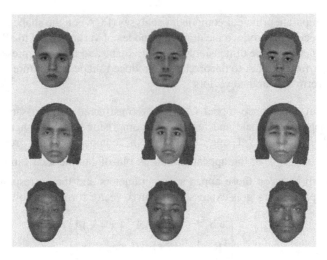

Fig. 1. Examples of randomly generated faces from the white male, Indian female and black male sub-groups

3   Interactive Evolutionary Face Search

In this section, we describe a stochastic search procedure for the genotype from which a likeness to the target face can be constructed as described by equations (1) – (3). It is crucial to recognize that the optimum search procedure for this task must be an algorithm that is a suitable compromise between *human usability* and *speed of convergence*. In this context, speed of convergence is properly defined as the required number of faces seen and rated by the witness before a satisfactory composite is achieved and is thus quite distinct from the computer execution time. Evolutionary algorithms [9] can be easily adapted to accommodate different forms of interactive input from a witness, and hence are well suited to an optimization problem of this type. Accordingly, an evolutionary algorithm (referred to as the "Select, Mutate and Multiply" or SMM algorithm) was designed to determine an optimal set of appearance model parameters. The SMM algorithm satisfies the requirement of simplicity and shows good convergence properties (see [10] for full details of its design and performance). The process by which the SMM algorithm generates a facial likeness is illustrated in figure 2 and can be summarized as follows –

- ❖ An initial population of nine randomly generated genotypes are converted to their phenotypes using equations (1) – (3) and presented in a 3 x 3 array to the witness via a graphical user interface.
- ❖ The witness selects the phenotype that exhibits the best overall likeness to the suspect's face (the preferred face). We note for future reference (section 4), that the witness may optionally identify and reject one or more of the poorest faces in the generation. We refer to these as the *reject(s)*.
- ❖ The corresponding genotype of the preferred face is then repeatedly cloned and each copy *mutated* by randomly perturbing its genes (i.e. appearance pa-

rameter values) with a certain (dynamic) probability. Each mutated clone will only be converted to its phenotype with a specified conversion probability β. By default $\beta = 1$, but we may set $\beta < 1$ in a way which is determined by the history of selected and rejected faces (see section 4)

❖ The preferred genotype and eight mutated genotypes form a new generation in which their corresponding phenotypes are randomly positioned and displayed to the witness.

❖ These steps are repeated until an acceptable likeness to the suspect's face is achieved.

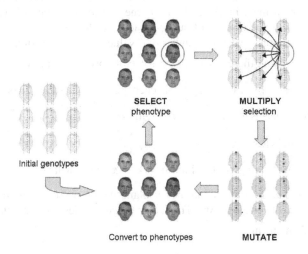

Fig. 2. The Select-multiply-mutate (SMM) evolutionary algorithm

The mutation rate determines the probability of each appearance parameter changing its value and is controlled dynamically according to the expression –

$$p(t) = 0.1 + 0.417\, t^{-0.558} \tag{5}$$

where $p(t)$ is the probability that a parameter will mutate and t is the number of generations produced up to and including the current generation. The parameters and form of equation (5) were optimized by computer simulated trials that mimicked the response of a human witness to the facial stimuli [10]. In this way, the degree of genetic modification is progressively reduced as the facial likeness improves, thereby producing a faster convergence to the suspect's face.

4 Improving the Efficiency of the Search Algorithm

In section 3 above, we described the basic SMM algorithm which is employed in the commercial incarnation of the system [4]. The SMM algorithm has been found empirically to establish a good compromise between simplicity of operation (a human witness can easily make the required decisions) and fast convergence properties.

Here, we now draw explicit attention to the fact that our implementation of the SMM algorithm provides the opportunity for a witness to explicitly *reject* one or more faces in each generation. The ability to reject weak members in each generation appears to have some psychological benefit to witnesses as it is generally easier for the witness to choose the worst examples and thereby simplify the task of selecting the preferred face. However, to date, no explicit reference has been made to the use of rejected faces in the algorithm to speed convergence. It is clear that the rejection of certain faces provides definite information (i.e. "don't show that face again or one that it close to it") which may, in principle, be harnessed to accelerate convergence to a satisfactory solution. The essence of our approach is as follows.

Candidate solution vectors are real-valued, N-dimensional genotypes $\mathbf{x} = \left[x_1, x_2 \cdots, x_N \right]$ and there exists a target vector corresponding to the 'ideal' solution denoted by $\mathbf{x}^T = \left[x_1^T, x_2^T \cdots, x_N^T \right]$. In each generation, the witness identifies a preferred vector which is *selected* by the witness $\mathbf{x}^S = \left[x_1^S, x_2^S \cdots, x_N^S \right]$ and a least preferred or *rejected* vector $\mathbf{x}^R = \left[x_1^R, x_2^R \cdots, x_N^R \right]$.

Consider a point P lying on the line $\mathbf{w} = \mathbf{x}^S - \mathbf{x}^R$ and distance $\alpha |\mathbf{w}|$ from point $\mathbf{x}^R$. We construct an N-dimensional hyper-plane which passes through point P and which is orthogonal to the line $\mathbf{w}$. The hyper-plane defines a discriminant function $g(\mathbf{x})$ which has the form –

$$g(\mathbf{x}) = \mathbf{W} \cdot \mathbf{x} + \omega_0 = 0 \tag{6}$$

$$\mathbf{W} = \mathbf{x}_S - \mathbf{x}_R, \quad \omega_0 = -\left(\mathbf{W} \cdot \mathbf{x}_R + \alpha |\mathbf{W}|^2 \right) \tag{7}$$

The discriminant function divides the space into two mutually exclusive regions - $\mathbb{R}_S$ the region in which $\mathbf{x}^S$ is located and $\mathbb{R}_R$ the region in which $\mathbf{x}^R$ is located. In general, for an arbitrary genotype $\mathbf{x}$, we have –

$$\begin{aligned} g(\mathbf{x}) &> 0 \quad &\Rightarrow \quad &\mathbf{x} \in \mathbb{R}_S \\ g(\mathbf{x}) &\leq 0 \quad &\Rightarrow \quad &\mathbf{x} \in \mathbb{R}_R \end{aligned} \tag{8}$$

After we have constructed the discriminant function, any subsequent genotype produced by the SMM procedure which satisfies $g(\mathbf{x}) \leq 0$ has its phenotype conversion probability *correspondingly reduced* through multiplication by factor $p_R, 0 \leq p_R \leq 1$. In this way, the probability landscape is successively modified to favour generation from within those regions of the search space lying closer to $\mathbf{x}^S$. By the kth

generation, K discriminant functions $\{g_1(x'), g_2(x') \cdots g_K(x')\}$ have been produced such that an arbitrary genotype x has a conversion probability given by –

$$\beta(x) = (p_R)^m \quad \text{where } m = \sum_{k=1}^{K} \text{logical}\{g_k(x) > 0\} \tag{9}$$

The fundamental assumption in our approach is that the preferred face in the generation (genotype x^S) *will lie closer to the target vector* x^T than the rejected vector x^R. If this assumption were strictly valid, it would guarantee that the target face x^T *always lies* within $\mathbb{R}_S$. Under these circumstances, we would be justified in setting $\alpha = 1/2$ and $p_R = 0$ and thereby maximally reducing the volume of the search space at each step. However, we assert this only in the average, statistical sense and not absolutely. This is because the relationship between the perceptual similarity of two different faces to a given target and the Euclidean distance to that target in the model space is non-linear and it is possible that a face whose Euclidean distance is further from a target face than another may nonetheless be perceived as more similar.

Intuitively, it is reasonable to assume that progressively small (or even negative) values for α, which move the hyperplane closer to x^R, will progressively increase the likelihood that $g(x^T) > 0$ and thus that $x^T \in \mathbb{R}_S$. However, increasing α will generally reduce the volume of the search space which is partially suppressed. Accordingly, empirical optimisation takes place over a 2-D parameter space - α, which controls the position of the hyperplane and p_R, which controls the 'strength' at which the probability landscape is altered.

Finally, we note that in practice witnesses make three types of decision in the evolutionary process. These are *selection*, *explicit* rejection (in which a face is positively identified as a poor likeness and removed from view) and *implicit* rejection (in which the face is simply not chosen as the preferred likeness). In principle, we may also construct hyperplanes for *implicitly* rejected faces in which the probability landscape on the negative side of the hyperplane is weighted by some factor p_R^{imp} (in general, $p_R^{imp} < p_R$) but this method is not currently implemented.

5 Deterministic Changes to Facial Appearance

Providing the functionality for making deterministic changes to a composite image is crucial in allowing the flexibility to work with real witnesses and may be used to accelerate the convergence. For this reason, the EFIT-V system includes additional tools that allow the witness to make deterministic changes to the face in response to requests from the witness [4, 11]. Recently, we have implemented two additional functions which prove very useful in practice and these are briefly summarized.

5.1 Wrinkle-Maps – The Application of High-Frequency Details

EFIT-V easily and automatically produces aging effects using an holistic method which develops our earlier work on age progression [12] and is essentially a refinement of the facial prototyping technique described by Benson and Perrett [13]. In the following section, we describe how aging transformations can be enhanced using Fourier methods to enhance the fine facial details.

The essential idea is to extract fine facial details from a sample or 'donor' face **A** and to then apply these details to a 'recipient' composite face **B**. Let I_A be the image which corresponds to subject **A** and let I_d be an image containing the fine facial detail from I_A. If **A** is an elderly subject, then image I_d will be likely to contain significant facial wrinkles associated with the aging process and is thus termed a *wrinkle-map*. Wrinkle-maps can be obtained by passing image I_A through a high pass filter, or equivalently by subtracting a smoothed version of the image from itself as follows,

$$I_d(x, y) = I_A(x, y) - \left[I_A(x, y) ** s(x, y) \right] \tag{10}$$

where $s(x, y)$ is a smoothing kernel which is convolved to blur image I_A.

Once extracted from a `donor' subject, a wrinkle-map can be applied to a composite face thereby enhancing its fine detail content and increasing the apparent age of a subject according to the equation -

$$I_B^* = (\gamma - 1)\left[I_B(x, y) ** s(x, y) \right] + I_d \tag{11}$$

In the normal implementation of high-boost filtering, a detail image is constructed as defined by equation (10) and then added to a specified fraction of the *original* image. In our case, the detail map is extracted from a different subject (the donor) which is then added to the subject (the recipient) and can thus be considered as a hybrid form of the high-boost filtering method. One of the advantages of this approach is that a wide variety of wrinkle maps may be applied to a composite image. The accurate registration of the wrinkle maps between donor and recipient images is also ensured by first applying a piecewise affine warp to the images to ensure they are both referred to a shape-normalized system. This basic procedure is illustrated schematically in figure 3.

5.2 Dynamic Overlays

The ability to add significant details and subtle refinements to a basic composite image form a very important aspect of day-to-day work for a composite operator. For this reason, practically all commercial composite systems allow access to industry standard paint programs to effect the necessary refinements. The sum total of all such external refinements is commonly termed the 'paint layer'. To the best of our knowledge, all other existing commercial systems treat the paint layer as *static* with respect to the composite image. A consequence is that adjustment of the composite once paint has been applied can be difficult and time-consuming for the operator.

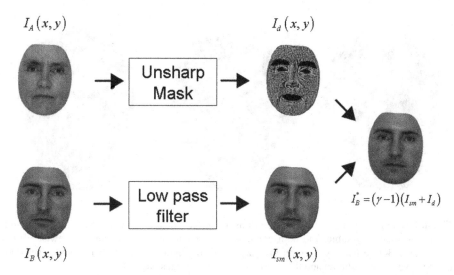

Fig. 3. The application of wrinkle-maps to enhance the effects of aging. An unsharp mask is applied to a donor face to extract the high frequency content. This is then added to a smoothed version of the recipient face.

We have developed a method in which paint-layers created by an artist to achieve a certain desired effect (either during the composite construction or beforehand) may be loaded onto the composite image and then automatically maintain registration with the underlying composite. In this way, a wide variety of artistic effects and additions may be pre-stored and loaded onto the composite, allowing considerable flexibility in operation.

Consider the mean image possessing the average shape $\bar{\mathbf{x}}$ and the average texture $\bar{\mathbf{g}}$ over the training sample. This is exported to the paint program and a modified texture is created, $\mathbf{g}^*$. Since the shape coordinates of both the original and modified image are referred to the average shape, we can calculate a texture difference image (referred to the average shape coordinates) as –

$$\Delta\mathbf{g} = \mathbf{g}^* - \bar{\mathbf{g}} \tag{12}$$

The difference image is thus non-zero only in those regions of the image which have been altered with paint work. Consider now an arbitrary composite image with shape vector $\mathbf{x}_{cI}$ and texture $\mathbf{g}_{cI}$ and that $\mathbf{T}$ is the piecewise affine warp which transforms the average shape vector $\bar{\mathbf{x}}$ into $\mathbf{x}_{cI}$, $T\{\bar{\mathbf{x}}\} = \mathbf{x}_{cI}$. Some multiple of the difference image is added to the shape normalized texture–

$$\mathbf{g} = \mathbf{g}_{cI} + \alpha\Delta\mathbf{g} \tag{13}$$

and this combined texture is then subjected to the *same piecewise affine transform, T*. This procedure ensures that the generated paint work tracks subsequent changes in the shape of the composite image. An illustrative examples is given in figure 4.

Fig. 4. Dynamic overlays – artistic alterations and effects can be made to track the underlying composite image by subjecting both composite and paint layer to the same piecewise affine warp. Left: Before overlay. Centre: Overlays applied. Right: The face shape, nose, eyes and mouth have all been altered and the overlays track the changes naturally.

6 Summary and Conclusions

We have described the basis of a radically new approach to facial composite generation based on an holistic face model and evolutionary search methods. Complementary systematic methods for improving the realism of the composite image have been presented along with computational techniques which substantially increase the speed of convergence by modifying the probability landscape from which the faces are sampled. Some typical examples of composite images generated from the EFIT-V system are presented in figure 5.

Fig. 5. Examples of composite images produced using the EFIT-V system

In a controlled trial supported by the UK Home Office, one of the participating police forces reported a greater than 100% increase in useful intelligence gained as a result of using EFIT-V [14] and more than 50 licenses for the EFIT-V system are now in routine use by both UK and European police services. The new methodology also has potentially profound implications for the training of facial composite operators. For feature-based systems, considerable emphasis has been placed in the past on the need to train composite operators in cognitive interview techniques [15]. In fact, the cognitive interview typically constitutes as much as 50% of a composite operator's training. However, some research indicates that completing a recognition task such as that posed by our system immediately after a recall task can actually decrease the likeness of the composite to a suspect's face. Some researchers have already begun the process of exploring alternative interview techniques more suitable for recognition-based, facial composite systems [16].

References

1. Turner, J., Pike, G., Towell, N., Kemp, R., Bennett, P.: Making Faces: Comparing E-FIT construction techniques. In: Proceedings of The British Psychological Society, vol. 7, p. 78 (1999)
2. Solomon, C.J., Gibson, S.J., Pallares-Bejarano, A.: Photographic Quality Facial Composites for Criminal Investigations. In: The 10th Meeting of the International Association for Craniofacia Identification (2002)
3. Gibson, S.J., Solomon, C.J., Pallares-Bejarano, A., Maylin, M.I.S.: Synthesis of Photographic Quality Facial Composites using Evolutionary Algorithms. In: Proceedings of the British Machine Vision Conference, pp. 221–230 (2003)
4. See, http://www.visionmetric.com
5. Frowd, C.D., Hancock, P.J.B., Carson, D.: EvoFIT: A holistic, evolutionary facial imaging technique for creating composites. ACM Transactions on Applied Psychology (TAP) 1, 1–21 (2004)
6. Caldwell, C., Johnston, V.S.: Tracking a Criminal Through "Face-Space" with a Genetic Algorithm. In: Belew, R.K., Booker, L.B. (eds.) Proceedings of the Fourth International Conference on Genetic Algorithms, pp. 416–421. Morgan Kauffman Pubs., San Diego (1991)
7. Cootes, T., Taylor, C.: Statistical models of appearance for medical image analysis and computer vision. In: SPIE Medical Imaging, pp. 236–248 (2001)
8. Solomon, C.J., Pallares-Bejarano, A., Gibson, S.J.: Non-Linear, Near Photo-Realistic Caricatures using a Parametric Facial Appearance Model. Behaviour Research: Methods, Instrumentation and Computing 37(2), 170–181 (2005)
9. Goldberg, D.E.: Genetic Algorithms in Search, Optimization and Machine Learning. Addison Wesley, Reading (1989)
10. Pallares-Bejarano, A.: Evolutionary Algorithms for Synthesis of Facial Composites, Ph.D thesis, University of Kent, UK (2006)
11. Gibson, S.J., Pallares-Bejarano, A., Maylin, M.I.S., Solomon, C.J.: The Generation of Facial Composites using an Evolutionary Algorithm. In: Procedings of 6th International Conference on Recent Advances in Soft Computing (best paper award), Canterbury (2006)
12. Scandrett, C.M., Solomon, C.J., Gibson, S.J.: A Person-Specific, Rigorous Aging Model of the Human Face. Pattern Recognition Letters 27(15), 1776–1787 (2006)
13. Benson, P.J., Perrett, D.I.: Extracting Prototypical Facial Images from Exemplars. Perception 22, 257–262 (1993)
14. Police Professional, Investigative Practice Journal, 21-24 (January 24, 2008)
15. Fisher, R.P., Geiselman, R.E.: Memory Enhancing Techniques for Investigative Interviewing. Charles Thomas, Springfield (1992)
16. Frowd, C.D., Bruce, V., Smith, A., Hancock, P.J.B.: Improving the quality of facial composites using a holistic cognitive interview. Journal of Experimental Psychology: Applied 14, 276–287 (2008)

Geovisualization Approaches for Spatio-temporal Crime Scene Analysis – Towards 4D Crime Mapping

Markus Wolff and Hartmut Asche

University of Potsdam, Department of Geography
Karl-Liebknecht-Strasse 24/25, 14476 Potsdam, Germany
{Markus.Wolff,gislab}@uni-potsdam.de

Abstract. This paper presents a set of methods and techniques for analysis and multidimensional visualisation of crime scenes in a German city. As a first step the approach implies spatio-temporal analysis of crime scenes. Against this background a GIS-based application is developed that facilitates discovering initial trends in spatio-temporal crime scene distributions even for a GIS untrained user. Based on these results further spatio-temporal analysis is conducted to detect variations of certain hotspots in space and time. In a next step these findings of crime scene analysis are integrated into a geovirtual environment. Behind this background the concept of the space-time cube is adopted to allow for visual analysis of repeat burglary victimisation. Since these procedures require incorporating temporal elements into virtual 3D environments, basic methods for 4D crime scene visualisation are outlined in this paper.

Keywords: 3D crime mapping, geovirtual environments, GIS, spatio-temporal.

1 Introduction

In contrast to traditional crime mapping analysis which results predominantly in two-dimensional static maps, the presented approach allows for creating interactive four-dimensional map applications. Therefore the paper focuses on combining methods of geospatial crime scene analysis with innovative spatio-temporal (4D) geovisualization techniques. As this concept links digital processing and geospatial analysis of crime data with easy-to-comprehend 3D and 4D visualisations respectively, the GIS (geographic information system) and VIS (visualisation system) tasks are combined in a specific workflow designed for that purpose.

Subsequent to a brief review of relevant related work (Section 2), initial spatio-temporal exploration of crime scene data is conducted (Section 3.1). For that purpose a GIS-based application is programmed that focuses on straightforward spatio-temporal crime scene exploration. To make this application as easy operable as possible – even for non-GIS specialists – the tool is designed as a standalone application using Esri's ArcGIS Engine development framework. Since its functionality focuses on initial exploration of temporal and spatial trends in crime scene distributions, the user does neither have to be familiar with ArcGIS nor with any other GIS.

The findings of this analysis are the starting point for further in-depth analysis (Section 3.2). Identified hotspots of certain crimes are analysed more detailed regarding

Z.J.M.H. Geradts, K.Y. Franke, and C.J. Veenman (Eds.): IWCF 2009, LNCS 5718, pp. 78–89, 2009.

their variation in space and time. Against this background a method is presented to visualise repeat residential burglary victimisation. Another visualisation method gives insights into temporal patterns regarding robbery offences by using 4D visualisation techniques. To compile advanced visualisations, analysis results are pipelined to a sophisticated 3D visualisation system (Autodesk LandXplorer Studio Professional). Depending on the analysis task, this system is used to design distinct 3D and 4D visualisations.

2 Related Work

This section provides a brief overview of existing studies in the disciplines of crime mapping and 3D geovisualization.

2.1 Crime Mapping

Among the various publications addressing crime mapping applications in theory and practical application [1] give an introduction into theories, methods and selected software systems used to document, monitor and analyse crime data. Because most crime analyses are based on geocoded crime scene data, precise geocoding is a major prerequisite for spatial analysis. Reference [2] reviews this topic of spatial data (in-) accuracy. However, dealing with geospatial crime scene data, an eminent task is to detect and map spatial hotspots of certain offences. According to [3] cited in [4], a hotspot is defined as an "area with high crime intensity". In addition to [1], an introduction into the different approaches of detecting and mapping hotspots can be found in [5]. With "prospective hotspot mapping" a different approach is introduced by [6]. However, areas with higher crime rates than other areas should be tested statistically. Reference [7] provides some insights addressing issues of statistical test and functions. Reference [8] finally gives both an overview of existing studies in the field of spatio-temporal crime mapping and applicable methods and techniques for visualising space and time related issues in crime patterns.

2.2 3D Geovisualization

Key themes and issues of geovisualization are presented by [9]. Based on their conceptual framework [10] discuss a research agenda on cognitive and usability issues in geovisualization. Adapting techniques of information visualisation to the requirements of cartography, [11] highlights that modern geovisualization helps to "stimulate visual thinking about geospatial patterns, relationships and trends". The author furthermore emphasises the advantage of creating three-dimensional visualisations that allow for an "additional variable to be displayed in a single view and, as such, gives the user a direct insight into the relationship between these variables". Reference [12] argues that the user can easily interpret spatial relationships of three-dimensional presented geo-objects without having to consult a legend. Reference [13] highlights the fact that three-dimensional cartographic visualisations provide "a more intuitive acquisition of space, due to an explicit use of 3D". Both, [12] and [13] indicate also disadvantages concerning three-dimensional visualisations as for instance the absence of a single scale in perspective views, occlusion of objects, etc. However, construction of virtual three-dimensional geovirtual environments requires dedicated software systems. The adopted LandXplorer software can be considered as an appropriate system for visualising interactive three-dimensional maps and is introduced by [14] [15].

3 Using Geovirtual Environments for Visualising Spatio-temporal Analysis Findings

This section describes the application of the outlined research framework by using crime scene data of the German city of Cologne. Data representing robberies and residential burglaries during the year 2007 is provided by the police headquarters of the city of Cologne. Each crime scene is described as a single point object, geocoded by x- and y-coordinates. In addition to these coordinates each point carries further thematic attributes describing time of the offence. First of all, those crime scenes are analysed regarding their overall patterns of distributions in space and time (Section 3.1). Based on these findings further in-depth-analysis is conducted (Section 3.2). In a last step all results of spatio-temporal analysis are pipelined to a 3D visualisation system. This leads to 3D and 4D visualisation respectively. This combination of spatio-temporal crime scene analysis with methods from the field of 3D/4D geovisualization supports an instant grasp of complex spatial phenomena – primarily for decision makers in security agencies that are not predominantly trained to map reading and map interpretation.

3.1 Exploring Initial Spatio-temporal Characteristics

Prior to in-depth-analysis of certain crime hotspots a first step of analysis comprises the identification of overall spatio-temporal trends in crime scene datasets. To deliver this task from the burden of having to use a complex GIS, an application is programmed that allows untrained GIS users as well as GIS experts to explore fundamental spatio-temporal trends in the underlying crime data. Using the Esri ArcGIS Engine software development framework this application is developed as a standalone application (.exe). ArcGIS Engine is used for software development since this developer kit allows for an easy integration of (Arc-) GIS functions into custom applications via ArcObjects programming. However, since proprietary ArcGIS functionality is addressed, the end user needs the Esri ArcGIS Engine runtime library or a licensed ArcGIS to be installed at the local machine. Besides loading and viewing of geospatial crime scene datasets the application allows basically for two kinds of spatio-temporal analysis (cf. Figure 1):

- Rapid generation of overview charts depicting the number of offences during a specified period.
- Hotspot analysis by computing kernel-density-estimation (KDE) based surfaces of crime scene densities for specified periods.

The process of analysis presented in this paper starts with conducting overview analysis regarding the general temporal distributions of certain crimes (residential burglaries and robberies) in 2007. This initial spatio-temporal data exploration is conducted by using the application. In a first step the monthly sums of robberies and burglaries, respectively, are calculated using the applications temporal query functions (Figure 2).

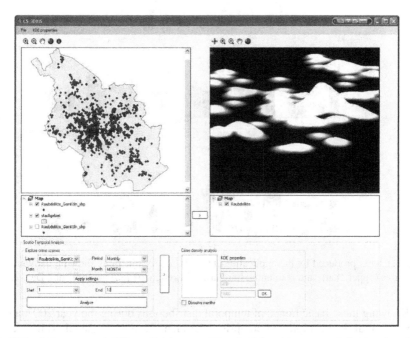

Fig. 1. GUI of programmed GIS-application providing initial spatio-temporal data exploration (CS-3DVIS.exe). While the left-hand window shows 2D map features, the right-hand window displays hotspot grids as three-dimensional surfaces.

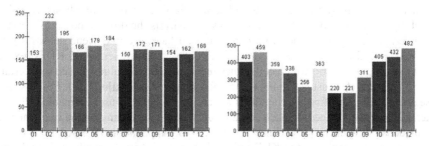

Fig. 2. Charts as produced by the applications temporal query functions

As Figure 2 shows, most robberies in 2007 are conducted during February (232 offences, left hand image), while most residential burglaries are committed in December (482 offences, right hand image). Against this background further analysis is conducted to exemplarily examine these two months more detailed. Therefore temporal resolution for analysis is increased by using the applications option "daily" for data exploration. This analysis reveals that most February robberies are conducted on Feb. 15th – which was Women's Carnival, the start of Cologne Carnival. Most residential burglaries in December on the other hand can be traced back to December 31st – New Year's Eve (cf. Figure 3).

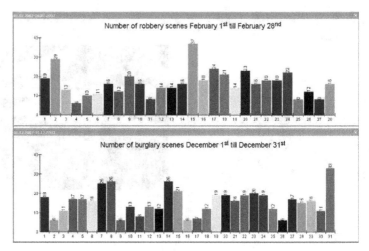

Fig. 3. Charts produced by the applications temporal query functions showing daily number of robberies (upper chart) and residential burglaries (lower chart) for Feb. 2007 and Dec. 2007

Knowing these basic trends of temporal distributions during the year 2007, a second step of initial data exploration comprises the identification of spatial crime scene distributions. For this purpose the applications KDE module is used. This supports a straightforward computation of kernel density surfaces for a selected dataset. To allow for further temporal analysis of these KDE surfaces, the application supports the calculation of KDE surfaces for specified periods. KDE surfaces can be created, for instance, on the basis of a year, of individual months or even on a daily basis. By using kernel density estimation techniques for hotspot analysis the discrete point distribution of crime scenes is transformed to a continuous surface of crime scene densities [16] [17] [18]. Based on the selected crime scene dataset, the implemented KDE module basically computes a grid which cell values represent density values related to a certain surface measure (for instance number of offences per square kilometer). For this purpose KDE-algorithms first overlay the study area with a grid of user definable cell size. Afterwards density values for each cell are calculated – depending on the implemented kernel density function [16]. Since the CS-3DVIS-application is based on ArcGIS functionality, the implemented KDE routine is based on ArcGIS quadratic kernel density function. To conduct KDE analysis, the user therefore has to specify the bandwidth and the output cell size for the KDE grid. Specification of bandwidth is crucial since this parameter describes the size of the search radius, i.e. how many crime scene locations (points) are used to calculate crime scene densities. Hence a large bandwidth includes a larger area and therefore more points into analysis than a smaller bandwidth would include. Therefore a too large bandwidth might hamper the identification of smaller hotspots, while a too small bandwidth might result in many small clusters of crime.

KDE-analysis carried out in this paper is based on a cell size of 20 meters and a bandwidth of 400 meters. The decision for the 400 meter search radius is taken predominantly as the result of experimental studies: compared with other settings, the 400 meter parameterisation produces the most reasonable output since the resulting

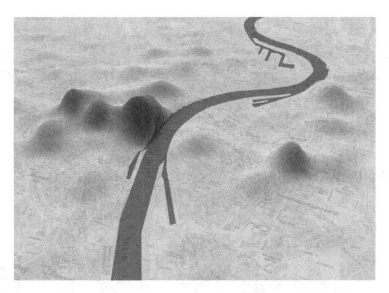

Fig. 4. 3D representation of a robbery hotspot surface for the year 2007

hotspot grid reveals very clearly inner-city hotspot-regions while preserves simulta-neously the overall representation of crime scene distribution. However, to allow for convenient bandwidth-experimentation the implemented KDE module supports multiple bandwidth specifications. To create for instance, multiple KDE surfaces with bandwidths of 50, 100, 150, 200, 250, 300, 350 and 400 meter, the user may enter these different values separated by a ";" into the corresponding bandwidth-specification field. Afterwards (visual) analysis of the created grids can help to identify the appropriate bandwidth setting.

Using the KDE module (applied bandwidth = 400 m, cell size = 20m) robbery scene densities are calculated initially for the year 2007 and in a second step for each month individually. After post processing (classification and colouring) these grids are pipelined to the 3D VIS LandXplorer and combined with other geocoded textures (as for instance with a aerial photography or a topographic map, cf. Figure 4). This multiple feature coding of crime scene densities can be considered as an effective visualisation method to single out certain hotspot regions and to quickly brief decision makers in security agencies for instance.

Because Figure 2 reveals a temporal variation of robbery offences during the year, the monthly distributions of crime scenes are analysed in a next step. For this purpose hotspot grids are computed for each individual month. The analysis reveals that those robbery hotspots in the city of Cologne vary clearly in time and space. Characteristic hotspots for instance are detected in February and August (Cologne Carnival and summer holidays, respectively). Afterwards each grid is pipelined to the 3D VIS and 12 images are generated from an identical point of view in a next step. Based on this images an animated Adobe Flash movie-sequence is produced, that visualises varying hotspot patterns in space and time. This animated sequence expands the set of spatio-temporal visualisation methods that help communicating complex geospatial phenomena to decision makers in an intuitive way.

Fig. 5. 3D representation of a residential burglary hotspot surface for the year 2007

Besides robberies also residential burglary crime scenes are analysed for spatial clustering by conducting KDE analysis. Unlike parameterisation for robbery analysis, KDE bandwidth for burglaries is specified by a value of 800 meters. Figure 5 depicts the corresponding hotspot distribution for residential burglary crime scenes.

3.2 Methods of 4D Spatio-temporal Crime Scene Visualisation

In the previous section methods for discovering general trends in the spatio-temporal distribution of crime scenes were presented as well as selected 3D visualisation methods. This section describes additional techniques for further time related analysis and 3D/4D visualisation, respectively. Since previous analysis revealed February 15th as the day with the highest number of robberies in 2007, further in depth-analysis is applied to this specific date. Therefore the resolution of temporal analysis is increased once more: exemplary for that day the offence's time of day is used as a focal point for further analysis.

Regarding this high-resolution spatio-temporal analysis the concept of the space-time cube is applied. In 1970 Hägerstrand introduced his fundamental concept of the so-called space-time cube [19] which is up to now adopted in a number of studies and applications (e.g. [20] [21])

Taking this concept into account the chronology of robberies is visualised using 4D geovisualization techniques. Taking the offences time of day as additional information, the vertical axis of the environment is used for representing time (cf. Figure 6). Thus, the Figure depicts two issues: the spatial distribution of robberies during February 15th in the city of Cologne (visualised by poles) and the temporal chronology of robberies, also known as space-time-paths (visualised as colored line segments that

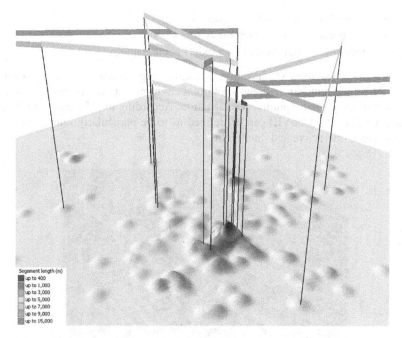

Fig. 6. Representation of space-time-paths for robberies on Feb. 15th in the city of Cologne

connect the robberies). The greater the z-values (the higher the line is above ground), the later on February 15th the offence occurred. The longer the space-time-paths, the more time elapsed between two robberies.

This kind of visualisation allows for an interactive exploration of complex temporal issues in geospatial crime scene analysis since it provides information about possible interrelations between certain offences. A short line segment between two offences, for example, may point to a possible link between these robberies (the Figure shows two such relatively short segments in red colours). Since such offences are situated closely together in time and space they may be committed by the same offender. However, this is a first assumption that has to be verified by taking further offence-related information into account. For such detailed geoprofiling additional police information is necessary. For instance time and space between two offences could be compared to the road- and the public transport network.

Another application potential of the space-time cube concept is the 4D visualisation of repeat victimisation. According to [22] repeat victimisation can be defined as "where the same offence occurs a number of times against the same victim (be this a person or an entity like a house)". However, [23] cited in [24] emphasise that repeats can also be "unconnected" with the initial occurrence. Therefore an additional specification to the definition above is "connection", i.e. if two burglaries, for instance, occur not just at the same building but also with the same employed modus operandi [23]. Since this definition contains

- spatial (where does the offence occur?)
- temporal (when and how many times?) and
- thematic (which modus operandi?) phenomena

the design of an intuitive visualisation is challenging. Given the multidimensional facet of this issue, multidimensional geovirtual environments can be considered as an adequate tool to facilitate intuitive visualisation. According to [25], cited in [26] geo-virtual environments (GeoVE) can be defined as to be particularly immersive, information intense, interactive and intelligent.

Fig. 7. 3D GeoVE with 3D city model, aerial photography and crime scenes

Therefore an extensive geovirtual environment for the city of Cologne is created. This consists of a digital terrain model, a 3D city model, high resolution aerial photography (25 cm/pixel), digital cadastral map and further vector-based datasets including rivers, administrative boundaries and others. Figure 7 shows exemplary a GeoVE including an aerial photography, a 3D city model and the position of crime scenes (burglaries 2007).

To allow for visual analysis of repeat residential burglary victimisation, time related information is integrated into the geovirtual environment. Since Figure 5 shows two fundamental hotspots of residential burglary (one located eastern, the other south-eastern of Cologne city centre), the crime scenes within the south-eastern hotspot are exemplarily used for further geovisual analysis. In this context the time of the burglary as well as the applied modus operandi is subject for analysis and visualisation (cf. Figure 8).

Each symbol in Figure 8 represents the position of a single residential burglary in the city of Cologne and a pole points to the burgled 3D building. The applied modus operandi is illustrated by a symbol representing the modus operandi "front door", "terrace door", "French window", "window" and "not registered". These symbols are stacked vertically according to the working day the burglary is committed. The

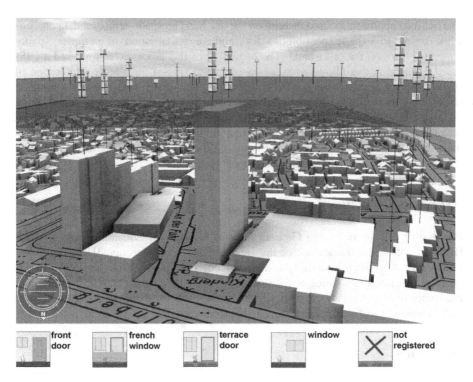

| | front door | | french window | | terrace door | | window | | not registered |

Fig. 8. Repeat residential burglary victimisation and applied modus operandi visualised in a geovirtual environment. Temporal dimension is visualised using the z-axis by classifying burglary scenes regarding the weekday of perpetration.

symbol at the lower end of the pole represents a burglary committed on a Monday, the second lowest stands for a Tuesday and so on, while the symbol on the top of the pole represents a burglary committed on a Sunday. To alleviate readability and comparability of different stacks, an optional layer (cf. dark blue transparent layer in Figure 8) is used to mark the temporal basement (that is a Monday).

Using this interactive visualisation, a crime analyst is allowed to explore three entities at the same time: 1st the spatial distribution of burglary offences within the study area, 2nd the temporal resolution and 3rd the employed modus operandi per burgled building. The combined visual analysis of all three elements allows for a convenient evaluation of repeat victimisation and makes this visualisation method and appropriate one for visual analysis tasks. Since the Figures are presented here as static screenshot images, it has to be underlined however, that the environment is completely interactive.

4   Conclusion

This paper demonstrated a set of methods to combine geospatial crime scene analysis techniques with the multidimensional visualisation potential of geovirtual environments.

The outlined approach focused thereby on the one hand on spatio-temporal data analysis while trying to explore possibilities for intuitive visualisation of complex spatio-temporal issues on the other hand. The beginning of this process chain marked a GIS-based application that allows even a GIS untrained analyst for a basic exploration of spatio-temporal features of geocoded crime scenes. Subsequent to this analysis, hotspots respectively their variations in space and time were analysed. Afterwards, the resulting files were pipelined to a 3D visualisation system. To put these results into an original urban context, a three-dimensional geovirtual environment was created. This environment was enhanced by the concept of the space-time cube for the purpose of integrating temporal issues of crime as, for instance, repeat victimisation.

Acknowledgements

Funding of this study by the German Federal Ministry of Education and Research (BMBF) within the framework of the InnoProfile research group '3D Geoinformation' (www.3dgi.de) is gratefully acknowledged. The author thanks the police headquarters of the city of Cologne for providing extensive crime datasets. Furthermore the author likes to thank Virtual City Systems GmbH for providing the 3D city model and Autodesk, Inc. for supplying the LandXplorer system.

References

1. Chainey, S., Ratcliffe, J.: GIS and Crime Mapping. John Wiley & Sons Inc, Chichester (2005)
2. McCarthy, T., Ratcliffe, J.: Garbage in, garbage out: geocoding accuracy and spatial analysis of crime. In: Wang, F. (ed.) Geographic Information Systems and Crime Analysis, IGI Global (2005)
3. Ratcliffe, J.H.: The Hotspot Matrix: A Framework for the Spatio-Temporal Targeting of Crime Reduction, pp. 5–23. Routledge, New York (2004)
4. Boba, R.L.: Crime Analysis and Crime Mapping. Sage Publications, Thousand Oaks (2005)
5. McCullagh, M.J.: Detecting Hotspots in Time and Space. In: ISG 2006 (2006)
6. Bowers, K.J., Johnson, S.D., Pease, K.: Prospective Hot-Spotting The Future of Crime Mapping. The British Journal of Criminology 44(5), 641–658 (2004)
7. Craglia, M., Haining, R., Wiles, P.: A Comparative Evaluation of Approaches to Urban Crime Pattern Analysis. Urban Studies 37(4), 711–729 (2000)
8. Brunsdon, C., Corcoran, J., Higgs, G.: Visualising space and time in crime patterns: A comparison of methods. Computers, Environment and urban Systems 31(1), 52–75 (2007)
9. MacEachren, A.M., Kraak, M.J.: Research challenges in geovisualization. Cartography and Geographic Information Science 28(1), 3–12 (2001)
10. Slocum, T.A., Blok, C., Jiang, B., Koussoulakou, A., Montello, D.R., Fuhrmann, S., Hedley, N.R.: Cognitive and Usability Issues in Geovisualization. Cartography and Geographic Information Science 28(1), 61–75 (2001)
11. Kraak, M.J.: Current trends in visualization of geospatial data with special reference to cartography. In: Indian Cartographer SDI-01, pp. 319–324 (2002)
12. Meng, L.: How can 3D geovisualization please users' eyes better. Geoinformatics - Magazine for Geo-IT Professionals 5, 34–35 (2002)

13. Jobst, M., Germanichs, T.: The Employment of 3D in Cartography - An Overview. In: Cartwright, W., Peterson, M.P., Gartner, G. (eds.) Multimedia Cartography, pp. 217–228. Springer, Heidelberg (2007)
14. Doellner, J., Baumann, K., Buchholz, H.: Virtual 3D City Models as Foundation of Complex Urban Information Spaces. In: CORP, Vienna (2006)
15. Doellner, J., Baumann, K., Kersting, O.: LandExplorer–Ein System für interaktive 3D-Karten. Kartographische Schriften 7, 67–76 (2003)
16. de Smith, M.J., Goodchild, M.F., Longley, P.A.: Geospatial Analysis. Troubador Publishing (2006)
17. Williamson, D., McLafferty, S., McGuire, P., Ross, T., Mollenkopf, J., Goldsmith, V., Quinn, S.: Tools in the spatial analysis of crime. In: Hirschfield, A., Bowers, K. (eds.) Mapping and analysing crime data. Taylor & Francis, London (2001)
18. Danese, M., Lazzari, M., Murgante, B.: Kernel Density Estimation Methods for a Geostatistical Approach in Seismic Risk Analysis: The Case Study of PotenzaHilltopTown (Southern Italy). In: Proceedings of International Conference on Computational Science and Its Applications (Part I). Section: Workshop on Geographical Analysis, Urban Modeling, Spatial Statistics (GEO-AN-MOD 2008), pp. 415–429. Springer, Heidelberg (2008)
19. Hägerstrand, T.: What about People in Regional Science. Papers, Regional Science Association 24, 7–21 (1970)
20. Gatalsky, P., Andrienko, N., Andrienko, G.: Interactive analysis of event data using space-time cube. In: Eight International Conference on Information Visualisation (IV 2004). IEEE Computer Society, Los Alamitos (2004)
21. Kraak, M.J.: The space-time cube revisited from a geovisualization perspective. In: Proceedings of the 21st International Cartographic Conference (ICC), Durban, South Africa (2003)
22. Ashby, D., Craglia, M.: Profiling Places: Geodemographics and GIS. In: Newburn, T., Williamson, T., Wright, A. (eds.) Handbook of criminal investigation. Willan Publishing, UK (2007)
23. Ratcliffe, J., McCullagh, M.: Crime, Repeat Victimisation and GIS. In: Hirschfield, A., Bowers, K. (eds.) Mapping and analysing crime data, pp. 61–92. Taylor & Francis, London (2007)
24. Hirschfield, A., Bowers, K.: Mapping and analysing crime data. Taylor & Francis, London (2001)
25. MacEachren, A.M., Edsall, R., Haug, D., Baxter, R., Otto, G., Masters, R., Fuhrmann, S., Qian, L.: Virtual Environments for Geographic Visualization: Potential and Challenges. In: ACM Workshop on New Paradigms for Information Visualization and Manipulation, Kansas City. ACM, New York (1999)
26. Fuhrmann, S., MacEachren, A.M.: Navigation in Desktop Geovirtual Environments: Usability Assessment. In: Proceedings of the 20th ICA/ACI International Cartographic Conference, Beijing, China (August 06-10, 2001)

Multimedia Forensics Is Not Computer Forensics

Rainer Böhme[1], Felix C. Freiling[2], Thomas Gloe[1], and Matthias Kirchner[1]

[1] Technische Universität Dresden, Institute of Systems Architecture,
01062 Dresden, Germany
[2] University of Mannheim, Laboratory for Dependable Distributed Systems,
68131 Mannheim, Germany

Abstract. The recent popularity of research on topics of multimedia forensics justifies reflections on the definition of the field. This paper devises an ontology that structures forensic disciplines by their primary domain of evidence. In this sense, both multimedia forensics and computer forensics belong to the class of digital forensics, but they differ notably in the underlying observer model that defines the forensic investigator's view on (parts of) reality, which itself is not fully cognizable. Important consequences on the reliability of probative facts emerge with regard to available counter-forensic techniques: while perfect concealment of traces is possible for computer forensics, this level of certainty cannot be expected for manipulations of sensor data. We cite concrete examples and refer to established techniques to support our arguments.

1 Introduction

The advent of information and communication technology has created a digital revolution which is about to change our world fundamentally. Digital information stored in computing systems increasingly defines tangible parts of our lives and thereby becomes an ever larger part of reality. Moreover, many physical or 'real-world' social interactions are being replaced by their virtual counterparts through computer-mediated communication. As a consequence, the rule of law has to be extended to the digital sphere, including enforcement and prosecution of crimes. This raises the need to reconstruct, in a scientific and reliable way, sequences of actions performed in the digital sphere to find—or at least to approach—the truth about causal relationships. This is a prerequisite to hold potential perpetrators accountable for their actions and to deter imitators.

Endeavors to use scientific methods to gain probative facts in criminal investigations are referred to as forensic sciences (short: *forensics*). This term has its etymologic roots in the Latin word 'forum', which means 'main square', a place where public court hearings took place in ancient times. The term *computer forensics* has emerged to describe similar endeavors when computers are involved in criminal activities [1]. However, the definition of computer forensics is somewhat blurred, as computers can stand in manifold relations to crimes: they can be tools to commit crimes in the real world, or means that merely create a digital sphere in which crimes take place. In both cases, forensic investigators may strive to extract probative facts from the computers involved.

Z.J.M.H. Geradts, K.Y. Franke, and C.J. Veenman (Eds.): IWCF 2009, LNCS 5718, pp. 90–103, 2009.

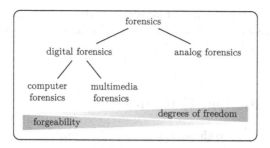

Fig. 1. Ontology of forensics, digital forensics and multimedia forensics

The situation becomes even more complex when we introduce *sensors* to the scenario. Sensors can capture parts of the reality and transform them into digital representations, such as images or audio files, which are then stored and processed in computers. Such digital representations of parts of reality can then be subject to forensic investigations, but they can only serve as probative facts if they are reliable and authentic. Realizing this goal defines the field of *multimedia forensics*.

This paper strives to clarify the definition of the various new variants of forensics, and to reflect on their underlying assumptions in comparison to classical forensic sciences known from the analog world. To do so, we employ an ontology of terms as illustrated in Fig. 1. One can subdivide all forensic sciences by their *domain of evidence*. This is the domain from which facts are extracted: classical (analog) forensics sets out to find traces of *physical evidence*, whereas *digital forensics* is limited to explore *digital evidence* [2]. While most people have a good intuition about the various forms of evidence derived from physical matters, digital evidence is intangible and therefore appears more abstract. Whenever we speak of digital evidence, we mean finite sequences of discrete and perfectly observable symbols, typically drawn from a binary alphabet, such as bit strings extracted from a computer's memory and storage devices. So both *computer forensics* and *multimedia forensics* share their reliance on digital evidence and thus can broadly be subsumed to digital forensics. In the following, however, we argue that they differ substantially in their underlying assumptions, which justifies the distinction made in the title of this work. We would like to point out that we intentionally draw a very black-and-white picture of the addressed sub-disciplines in order to highlight their basic differences. Many practitioners from the one or the other field may be inclined to disagree with some of the assertions made. In practical investigations, of course, we will see a more grayish picture with combinations of different disciplines. We believe that such combinations in practice blur the important differences. This calls for a structured approach, to which this paper shall make a novel contribution.

In the remainder of this paper, we discuss each branch in more detail, taking classical (analog) forensics as a starting point (Sect. 2). Adhering to our terminology, we recall the principles of computer forensics in Section 3 to distinguish it from the discipline of multimedia forensics in Section 4. In Section 5 we change the perspective to counter-forensics and discuss the main challenges for each of the two branches of digital forensics. The final Section 6 concludes with remarks on the possibility to combine the various sub-disciplines in practical investigations.

2   Classical (Analog) Forensics

Classical forensics refers to the endeavor to extract probative facts from physical evidence in the reality, i. e., the 'analog' world. It has been argued that the discipline draws on two principles: (a) *divisibility of matter* [3], and (b) *transfer* [4,3]. The first principle means that matter divides into smaller parts when sufficient force is applied. The smaller parts retain characteristics of the original matter as well as acquire characteristics generated by the separation itself.

The second principle, also known as *exchange principle*, states that whenever two entities interact in the real world, e. g., a burglar and a padlock, each entity will retain some physical matter of the other [4,5,6, among others]. Such exchanges can include for example fingerprints and footprints, hair, fibres of clothes, scratches, wounds, or oil stains. The examples show that transfer should not only be reduced to transfer on a microscopic scale. As Inman and Rudin [3] emphasize, transfer also includes the exchange of *patterns* (like footprints). So transfer means not only transfer of matter, but also transfer of traits.

If one accepts the principles as given, then it is straight to follow Kirk [7]:

> "Physical evidence cannot be wrong, it cannot perjure itself, it cannot be wholly absent. Only human failure to find it, study and understand it, can diminish its value."

This means that an *unconstrained* forensic investigator—at least theoretically—is free to analyse the scene (i. e., reality) from infinitely many perspectives (though not all at the same time). So he or she would have a non-zero chance to find even the subtlest trace. However, in modern epistemology it is accepted that human cognition of reality is in fact constrained in several ways. Most importantly, the human sense organs and the perceptual processes give us an incomplete picture of reality. While this can be considered as a filter which can be partly compensated for with technical means (e. g., microscopes increase the resolution of the human visual system), the Heisenberg uncertainty principle imposes an even greater constraint: the observer is always part of the very same reality and the sheer fact that he or she interacts with it, changes the object to be observed. This is also consistent with Inman and Rudin's reflections on the division and transfer of physical matter [6] which do not distinguish between perpetrators and forensic investigators as entities taking part in the exchange.

What is important about this view for the argument in this paper is the reliability of probative facts derived from physical evidence. In other words, how difficult is it for a very sophisticated perpetrator to wipe out all traces, or even worse, to forge traces that can lead to false accusations? This corresponds to the attempt to modify reality in order to create a different picture of his actions.

Since both forensic investigator and perpetrator are part of the same reality and therefore subject to similar physical and cognitive constraints, even the most sophisticated perpetrator can never be sure whether his 'modification of reality' is fully consistent with the (imaginary) reality if the action had not taken place. So committing a 'perfect crime' in reality, and pretending a consistent picture of reality that hides all traces, is an incredibly difficult problem. As a result, careful investigations of physical evidence are likely to deliver either reliable probative facts or none. (Ignoring the possibility of lapses, which can never be fully excluded.)

3   Computer Forensics

Computers are physical machines that form part of our reality. Therefore, at first sight, if one accepts the divisibility and transfer principles, then they should equally hold for computer forensics. However, when people speak of computer forensics, they often make the implicit assumption that the forensic analysis is limited to the digital evidence stored in the status of the finite automata each computer represents. This implies an observer model with drastically reduced view on reality: bits alone are theoretical concepts that carry no side-information about their history. For example, the common practice to make a copy of the digital evidence stored in a computer and to base further investigations exclusively on this (read-only) copy, implements this observer model [8,9]. This observer model also implies that in computer forensics, divisibility of matter is not relevant. The transfer of traits remains as a possible basis for a theory of computer forensics.

This is not without consequences on the reliability of probative facts derived from such digital evidence. As the number of states in a closed system is finite, there is *always* a non-negligible chance that a sophisticated perpetrator leaves a computer in a state which *perfectly* erases all traces. Assuming that the entire persistent state of a computer is stored on hard disk, this can be achieved, for example, by using the computer after booting from a live-CD (and not altering anything on the disk).

Perfectly erasing all traces in practice is of course not always easy. The number of possible states that need to be controlled quickly grows intractably high. For example, nowadays standard PCs are equipped with about 100 GB of disk space. This translates to about $2^{10^{11}}$ states; for comparison, the number of atoms in the universe is estimated in the order of magnitude of 2^{10^3}. Especially in the complex modern networked systems with their many software components and hardware interfaces, perpetrators often fail to control parts of the state space. However, the perpetrator could use more technology, e. g., another system that simulates the relevant computer at the crime scene in a virtual machine. This helps to construct a valid and plausible state with reasonable time and effort, as only a negligibly small fraction of all possible states is actually relevant for finding a clean system state. This, however, implies an observer model that only sees parts of the entire state space; the observer ignores the additional technology used to create the clean state. In practice, it is often hard for investigators to determine the borders of the system to be seized and analyzed, especially if it uses (wireless) network links.

Even with an observer model that captures the entire system, it follows from the limitation of the analysis to digital evidence that we can never ignore the possibility that a sophisticated perpetrator has covered all digital traces perfectly. In practice, such sophisticated perpetrators may be rare, but some skepticism is appropriate with regard to the residual probability of error whenever digital evidence is used in court to judge about capital crimes.

So does the principle of transfer apply to computer forensics? Many practicioners today will be inclined to agree, because from their experience every perpetrator makes mistakes and will leave patterns of criminal activity on the evidence. However, the digital nature of evidence makes it possible to cover traces perfectly. Furthermore, unlike practical limitations of the observer in classical (analog) forensics, the perpetrator knows all about this 'blind spot' of the investigator in advance and thus can adapt his action and pretend false or misleading facts. The advantages of inexpensive (due to automation) and convenient[1] computer forensics—most of the work can be carried out from the forensic investigator's office—come at the cost of lower probative force. As social interactions move into the digital sphere, state-funded investigation offices have to make delicate decisions on the allocation of resources between exploitation of physical and digital evidence.

A completely different situation emerges if computer forensics is understood in a broader sense that comprises both physical and digital evidence (unlike in this paper). Such additional physical evidence, although sometimes costly and cumbersome to obtain, can be very indicative side-information. Features such as wear and tear, recordings of electromagnetic emanations [10], temperature [11], as well as all kinds of analog traces on storage devices [12] might reveal information about previous states of a target computer and thus thwart efforts to conceal traces. Even digital (or digitized) evidence stored in other devices (e. g., computers connected over a network link) can form such additional information *if* their integrity is secured, e. g., by means of secure logging [13]. For example, US agent Oliver L. North was convicted in the Iran-Contra affair in 1986 because he had overlooked evidence that was stored on backup tapes.

4   Multimedia Forensics

An important class of digital data which is often found (and analyzed) on seized mass storage devices is digital multimedia data. While digital and digitized media nowadays affect (and mostly enrich) our everyday life in innumerable ways, critics have expressed concerns that it has never been so easy to manipulate media data. Sophisticated editing software enables even unexperienced users to

---

[1] Note that the largest inconvenience in modern digital investigations results from the enormous amounts of data on seized computers.

substantially alter digital media with only small effort and at high output quality. As a result, questions regarding media authenticity are of growing relevance and of particular interest in court, where consequential decisions might be based on evidence in the form of digital media.

Over the past couple of years, the relatively young field of multimedia forensics has grown dynamically and now brings together researcher from different communities, such as multimedia security, computer forensics, imaging, and signal processing. Although multimedia forensics, like computer forensics, is based on digital evidence, the fact that symbols are captured with a sensor makes a difference which has implications on the reliability of probative facts. In the following, we will briefly define the field and then discuss the relation to other forensic sciences.

4.1   A Short Introduction to Multimedia Forensics

Scholars in multimedia forensics aim at restoring some of the lost trustworthiness of digital media by developing tools to unveil conspicuous traces of previous manipulations, or to infer knowledge about the source device. We call these two basic branches of multimedia forensics *manipulation detection scenario* and *identification scenario*, respectively. Note that multimedia forensics in this sense is not about analyzing the semantics of digital or digitized media objects. Techniques from multimedia forensics merely provide a way to test for the authenticity and source of digital sensor data.[2] This is a prerequisite for further analysis: probative facts derived from the content of multimedia data (for instance speaker identification from a microphone recording or license plate identification from a CCTV video) are only useful when the underlying data is reliable and authentic.

In multimedia forensics, it is generally assumed that the forensic investigator does not have any knowledge of a presumed original. Such methods are called 'blind' [15] and typically exploit two main sources of digital traces:

▷ Characteristics of the acquisition device can be checked for their very presence (identification scenario) or consistence (manipulation detection scenario).
▷ Artifacts of previous processing operations can be detected in the manipulation detection scenario.

The first class of traces is inseparably connected with the process of capturing digital media [16]. Since different sensors systematically vary in the way they transform (parts of) reality into a discrete representation, each capturing device is believed to leave characteristic features in its output data. The level of variation determines whether the corresponding traces can be used to distinguish the class [17,18], model [19,20,21,22,23] or specific device [24,25,26] of an acquisition device. Today's multimedia forensic techniques mostly focus on the analysis of digital images. Here, one of the probably most-studied device characteristics is the CCD/CMOS sensor noise, which occurs in practically all

---

[2] A related discipline, which could be even framed into the general concept of multimedia forensics, is steganalysis. The link becomes obvious whenever we think of embedding a secret message as a manipulation of genuine sensor data [14].

Fig. 2. Typical image manipulation and detection with multimedia forensics. Presumably original photograph of Iranian missile test with one non-functioning missile (left, source: online service of the Iranian daily Jamejam today) which was replaced by a copy and paste forgery (middle, source: Iranian Revolutionary Guards). The detector output marks regions which were copied with high probability (right) [32].

digital cameras [25] or scanners [27]. Estimates of the so-called photo response non-uniformity (PRNU)—a noise source that reflects small but systematic deviations in the light sensitivity of single sensor elements—serve as 'digital fingerprints' that allow to identify individual acquisition devices. A useful analogy is the analysis of bullet scratches in classical forensics, which assign projectiles to weapons [28].

Besides the usefulness of such device-specific traces in the identification scenario, they also found wide application in the detection of manipulations [29,30,31,25]. By testing for the existence of consistent device characteristics in the whole digital media object, deviations from the genuine sensor output can be detected. For example, a block-by-block analysis that signals the absence of the expected PRNU in certain image regions can be seen as indication for possible (local) post-processing.

There are more ways to uncover manipulations of digital media. Traces of the applied post-processing itself can also be very indicative [32,33,34,35,36]. Forensic methods that exploit this type of traces approach the problem of manipulation detection from the opposite direction than techniques based on device characteristics. Not the absence, but the very presence of particular features is used as a probative fact for possible post-processing. Typical traces of manipulation include periodic inter-pixel correlations after geometric transformations, like scaling or rotation of digital images [33], or the identification of (almost-)duplicate regions after copy and paste operations [32]. An recent example for the latter type of manipulations is depicted in Fig. 2, which shows a forged image of an Iranian missile test that was analyzed with a copy and paste detector.

4.2   Relation to Computer Forensics

Even though both computer forensics and multimedia forensics explore digital evidence, we believe that they form two distinct sub-categories of digital forensics. This may seem counter-intuitive at first sight, since in any case, the domain of evidence is limited to the set of discrete symbols found on a particular device. In multimedia forensics, however, it is assumed that these discrete symbols were captured with some type of a sensor and therefore the symbols are a digital representation of an incognizable reality. The existence of a sensor that transforms natural phenomena to discrete projections, which are then subject to investigation, implies that multimedia forensics has to be seen as *empirical* science. This resembles the epistemological argument brought forward in the context of steganography in digitized covers [37]. Literally, a forensic investigator can never gain ultimate knowledge about whether a piece of digital media reflects reality or not. Neither can a sophisticated perpetrator be sure whether his manipulation really has not left any detectable traces. Unlike computer forensics, digital evidence in multimedia forensics is linked to the outside world and cannot be reproduced with machines. Thus, while the principle of transfer does not necessarily apply to computer forensics, it does have a place in multimedia forensics.

To make complex matters like a 'projection of reality to discrete symbols' more tractable with formal methods, multimedia forensics employs *models* of reality (though rarely stated explicitly). PRNU-based camera identification, for instance, assumes that the sensor noise follows some probability distribution, which can be reasonably approximated with a Gaussian distribution. This way, the problem can be formulated as a hypothesis testing problem with an optimal detector for the applied model [25]. Another sort of models is implied in methods that try to detect copy and paste operations. The assumption here is that connected regions of identical, but not constant, pixel values are very unlikely to occur in original images [32]. The two examples stress that typical models function as yet another dimensionality reduction within the domain of digital evidence. (A first reduction is the projection of physical evidence to digital evidence, see Sect. 1.) So the models provide a very simplistic view of reality.

Obviously, the quality of probative facts resulting from multimedia forensic methods depends on the quality of the model. The better an underlying model can explain and predict (details of) reality, the more confident we can base decisions on it. A model of PRNU which incorporates different image orientations is definitely preferable to one that does not. It can help to decrease the probability of missed detection. False alarms can be reduced by removing so-called non-unique artifacts like traces of demosaicing from the PRNU estimates [25].

It is important to note that the uncertainty about the generally incognizable reality is not the only fundamental difference between multimedia forensics and computer forensics. The transformation from analog world to discrete symbols itself adds further degrees of freedom on the sensor level. Especially the extent of *quantization* is a very influential factor for all multimedia forensic techniques, but in general every sort of post-processing inside the sensor has to be taken into account for a thorough analysis of digital media data. By definition, quantization causes information loss and thus introduces uncertainty in the forensic analysis. Here, quantization not only refers to lossy compression schemes like JPEG, but for instance also to the resolution of the output data.[3] When reasoning about what constitutes a *sensor* in a wider sense, i.e., including possible attached

---

[3] Note that JPEG compression is by far the most relevant source of uncertainty in practical applications: virtually all known forensic methods are more or less vulnerable to strong JPEG compression.

data compression mechanisms, sooner or later one stumbles over the question of 'legitimate' post-processing. For example, scans of printed and dithered images in newspapers can result only in a very coarse digital representation of reality, but traces of inconsistent lighting may still be detectable [36]. Generally, it appears that the quality of sensor output necessary for sound forensic analysis heavily depends on the applied techniques—yet another aspect which has no counterpart in computer forensics (in the narrow sense), where digital symbols are not linked to the world outside the closed and deterministic system.

Following our previous comments on computer forensics in a broader sense (cf. Sect. 3), we have to point out the general similarity of multimedia forensics with additional side-information about previous states. Strictly speaking, a computer becomes a sensor whenever it records signals of its environment, i. e., the reality. Recordings can happen for various reasons, for example key stroke pattern are used to seed pseudo-random generators. Such pattern also convey information about the typist, who no doubt belongs to the reality [38].

5   Counter-Forensics[4]

Theory and practices to reconstruct crime scenes and, hence, to identify evidence are not reserved to the special group of forensic investigators. Most state-of-the-art methods are published in publicly available conference proceedings or journal articles. In general, transparency is a welcome security principle [39], but at the same time it makes it a bit easier for potential perpetrators to refine their strategies and to develop counter-forensic methods, which reduce the formation or availability of probative facts to the forensics process [40].

The horizontal order of sub-disciplines in Fig. 1 has been chosen intentionally to reflect gradual differences in the reliability of probative facts. This is emphasized by the two scales on the bottom. The measure *degrees of freedom* captures the amount of possibilities through which an investigator—at least theoretically—is able to collect evidence from the scene. Obviously, it is highest for analog forensics (physical evidence) and lowest for computer forensics due to the restricted observer model. The more restricted and the better predictable the observer model is, the easier it becomes for sophisticated perpetrators to manipulate the facts undetectably. This can be expressed in a measure of *forgeability*, which forms a kind of mirror image to the degrees of freedom and directly relates to counter-forensics. So in the following, we will explain the order of sub-disciplines with respect to forgeability in more detail.

In classical forensics, and adhering to the principles of division and transfer, counter-forensic methods that completely avoid the formation of probative facts cannot exist. Even the most sophisticated perpetrator can merely compete with forensic investigators to find the best abstraction of the real world in order to include as many as possible traces in their actions. For example, a perpetrator

---

[4] The terms 'counter-forensics' and 'anti-forensics' appear synonymously in the literature. We prefer the former because it better reflects the *reaction to* forensics, as opposed to *disapproval of* forensics.

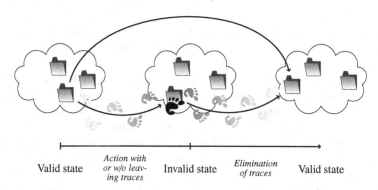

Fig. 3. Malicious actions cause traces by setting the machine to a suspiciously 'invalid' state that is detectable by forensics. Counter-forensics either eliminate all occurring traces subsequently (bottom path), or avoid traces preemptively (top path).

can remove traces like fingerprints by cleaning all touched surfaces. But indeed, cleaning surfaces with a cleanser and a cloth introduces new evidence. So the efforts to hide evidence transitively will most likely end up in an infinite recursion. Consequently, only human failure to find probative facts enables a perpetrator to be successful in the analog world (cf. Sect. 2).

This is totally different for computer forensics, where the discrete and finite nature of computer systems allows farsighted perpetrators, first, to determine valid states, and second, to reset a system from an invalid (i. e., suspicious) state back to a previously recorded valid state (see Fig. 3). For example, to cover data theft from a standard desktop PC, a perpetrator stores the initial state of the source device, transfers all portions of the demanded data, and finally, resets the source device to the initial state to remove occurring traces. While the success of this type of counter-forensic methods typically is limited to scenarios without any system inspection before the traces are eliminated, other counter-forensic methods can avoid detectable traces preemptively. A simple practical example is to boot an operating system from CD-ROM and mount the local hard drive in read-only mode. Figure 3 illustrates both approaches of counter-forensic methods.

Regarding multimedia forensics, we can distinguish two possible goals a perpetrator could strive for:

▷ altering the result of identification schemes by either suppressing the true source or counterfeiting a different one, and
▷ hiding post-processing by synthesis of authentic characteristics of the acquisition device or suppression of post-processing artifacts.

Practical examples for the first goal include attempts to suppress the device-specific noise pattern in a given image and replace it with the pattern of another camera [41,42]. Techniques to realize the second goal include a method for undetectable resampling [14] or the attempt to synthesize authentic demosaicing artifacts in arbitrary, possibly manipulated, images [43]. The two counter-forensic approaches sketched in Fig. 3 also apply to multimedia-forensics [14]: While generating a valid demosaicing pattern after a manipulation clearly aims at removing suspicious traces, employing a undetectable resampling approach is intended to never leave any traces at all.

However, eliminating or avoiding traces by setting a valid state is not so simple in the case of multimedia data. This is so because the discrete symbols, via the sensor, depend on the scene that is part of reality. Although the number of possible states is finite, unlike in the analog world, it is too large to determine valid states with reasonable effort. And in contrast to computer forensics in deterministic machines, it is impossible to escape this problem by simply 'virtualizing' reality in a larger system. Consequently, sophisticated perpetrators and forensic investigators compete for the best abstractions to model relations between digital data and real world scenes. Their goal is either to hide or to counterfeit digital evidence (perpetrator's point of view), or to detect even the subtlest modifications in media content (investigator's point of view).

The reason why practical counter-forensics happen to work in laboratory settings is that current forensic techniques base their decisions on very low-dimensional criteria. In other words, they rely on a very simplistic models of reality. It is unlikely that these counter-forensic methods will still be successful against a combination of a handful of forensic techniques, so that the dimensionality is somewhat higher. And it is an open research question whether models can be found good enough to fool such combinations with novel counter-forensic techniques. In the meantime, an alternative could be to discourage forensic analysis by increasing the uncertainty through lossy, but inconspicuous post-processing (information loss through lossy compression or size reduction). The focal point here is the question which post-processing will be perceived as inconspicuous. This seems to be an inverse problem to the question for legitimate post-processing in Sect. 4.2. Both answers ultimately depend on established habits and conventions, which themselves are conditional to context information and may change over time.

6   Concluding Remarks

In this paper, we have devised an ontology to structure the various kinds of forensic disciplines by their primary domain of evidence. We deem such a distinction appropriate to clarify the assumptions and the logic of inference behind the different sub-disciplines. In particular, it became evident that the fact whether digital evidence is collected from the real world with a sensor, or merely represents an internal state of a closed and deterministic system, makes a difference with respect to the reliability of the extracted probative facts: it is harder to forge media data undetectably than to manipulate other digital evidence. Further, the notion of an observer model helps to distinguish the two extremes—computer forensics and classical (analog) forensics.

One may rejoin that this distinction is fairly artificial, as our conceptual borders are quite blurred in practice. For example, a police search could result in a hard disk image, on which digital photographs are to be found with computer forensic methods. Then, multimedia forensics is applied to assign these photographs to a particular digital camera, which has been seized elsewhere. Fingerprints on this camera ultimately lead to the identity of the perpetrator via a police database. In this example, all kinds of forensic disciplines interact, spanning both digital and physical evidence, and jointly form a complete chain of evidence. While this holistic approach hopefully helps to convict the right person, such combinations in practice could hide the subtle differences between the various methods involved, and thus complicate the exercise to study each of them separately.

We see the contribution of this paper in a modest attempt to structure the field and to reflect on the (often implied) assumptions and models more explicitly and critically. Our proposal of an ontology and its accompanying terminology are understood as a starting point to stimulate fruitful discussion. Further refinements are envisaged for future research, along with an attempt to replace the informal arguments with more formal rigor. This implies that the deterministic view in this paper has to be replaced by the probabilistic theory of hypothesis testing.

Acknowledgements

The authors would like to thank the anonymous reviewers for their detailed and helpful comments and their colleague Stefan Köpsell for challenging our view of the world. Matthias Kirchner gratefully receives a doctorate scholarship from Deutsche Telekom Stiftung.

References

1. Kruse, W., Heiser, J.: Computer Forensics: Incident Response Essentials. Addison Wesley, Reading (2001)
2. Carrier, B., Spafford, E.H.: Getting physical with the digital investigation process. International Journal of Digital Evidence 2(2) (2003)
3. Inman, K., Rudin, N.: The origin of evidence. Forensic Science International 126, 11–16 (2002)
4. Locard, E.: L'Enquête criminelle et les Methodes scientifiques, Flammarion, Paris (1920)
5. Saferstein, R.: Criminalistics: An Introduction to Forensic Science, 7th edn. Prentice Hall, Englewood Cliffs (2000)
6. Inman, K., Rudin, N.: Principles and Practices of Criminalistics. CRC Press, Boca Raton (2000)
7. Kirk, P.L.: Crime Investigation. John Wiley & Sons Inc, Chichester (1974)
8. Casey, E.: Digital evidence and computer crime, 2nd edn. Academic Press, London (2004)
9. The Common Digital Evidence Storage Format Working Group: Standardizing digital evidence storage. Communications of the ACM 49(2), 67–68 (2006)
10. Kuhn, M.G.: Compromising emanations: eavesdropping risks of computer displays. PhD thesis, University of Cambridge Computer Laboratory (2003)

11. Zander, S., Murdoch, S.J.: An improved clock-skew measurement technique for revealing hidden services. In: SSYM 2008: Proceedings of the 17th USENIX Security Symposium. USENIX Association, Berkeley (2008)
12. Wright, C., Kleiman, D., Sundhar, S.: Overwriting hard drive data: The great wiping controversy. In: Sekar, R., Pujari, A.K. (eds.) ICISS 2008. LNCS, vol. 5352, pp. 243–257. Springer, Heidelberg (2008)
13. Schneier, B., Kelsey, J.: Cryptographic support for secure logs on untrusted machines. In: SSYM 1998: Proceedings of the 7th USENIX Security Symposium. USENIX Association, Berkeley (1998)
14. Kirchner, M., Böhme, R.: Hiding traces of resampling in digital images. IEEE Transactions on Information Forensics and Security 3(4), 582–592 (2008)
15. Ng, T.T., Chang, S.F., Lin, C.Y., Sun, Q.: Passive-blind image forensics. In: Zeng, W., Yu, H., Lin, C.Y. (eds.) Multimedia Security Technologies for Digital Rights, pp. 383–412. Academic Press, London (2006)
16. Khanna, N., Mikkilineni, A.K., Martone, A.F., Ali, G.N., Chiu, G.T.C., Allebach, J.P., Delp, E.J.: A survey of forensic characterization methods for physical devices. Digital Investigation 3(suppl. 1), 17–28 (2006)
17. Khanna, N., Chiu, G.T.C., Allebach, J.P., Delp, E.J.: Forensic techniques for classifying scanner, computer generated and digital camera images. In: Proceedings of the 2008 IEEE International Conference on Acoustics, Speech, and Signal Processing (ICASSP 2008), pp. 1653–1656 (2008)
18. McKay, C., Swaminathan, A., Gou, H., Wu, M.: Image acquisition forensics: Forensic analysis to identify imaging source. In: Proceedings of the 2008 IEEE International Conference on Acoustics, Speech, and Signal Processing (ICASSP 2008), pp. 1657–1660 (2008)
19. Kharrazi, M., Sencar, H.T., Memon, N.: Blind source camera identification. In: Proceedings of the 2004 IEEE International Conference on Image Processing (ICIP 2004), 709–712 (2004)
20. Böhme, R., Westfeld, A.: Feature-based encoder classification of compressed audio streams. Multimedia Systems Journal 11(2), 108–120 (2005)
21. Bayram, S., Sencar, H.T., Memon, N.: Classification of digital camera-models based on demosaicing artifacts. Digital Investigation 5, 46–59 (2008)
22. Farid, H.: Digital image ballistics from JPEG quantization: A followup study. Technical Report TR2008-638, Department of Computer Science, Dartmouth College, Hanover, NH, USA (2008)
23. Gloe, T., Borowka, K., Winkler, A.: Feature-based camera model identification works in practice: Results of a comprehensive evaluation study. In: Accepted for Information Hiding 2009, Darmstadt, Germany, June 7–10. LNCS (to appear, 2009)
24. Geradts, Z.J., Bijhold, J., Kieft, M., Kurosawa, K., Kuroki, K., Saitoh, N.: Methods for identification of images acquired with digital cameras. In: Bramble, S.K., Carapezza, E.M., Rudin, L.I. (eds.) Proceedings of SPIE: Enabling Technologies for Law Enforcement and Security, vol. 4232, pp. 505–512 (2001)
25. Chen, M., Fridrich, J., Goljan, M., Lukáš, J.: Determining image origin and integrity using sensor noise. IEEE Transactions on Information Forensics and Security 3(1), 74–90 (2008)
26. Dirik, A.E., Sencar, H.T., Memon, N.D.: Digital single lens reflex camera identification from traces of sensor dust. IEEE Transactions on Information Forensics and Security 3(3), 539–552 (2008)
27. Gloe, T., Franz, E., Winkler, A.: Forensics for flatbed scanners. In: Delp, E.J., Wong, P.W. (eds.) Proceedings of SPIE: Security and Watermarking of Multimedia Content IX, vol. 6505, p. 65051I (2007)

28. Lukáš, J., Fridrich, J., Goljan, M.: Digital "bullet scratches" for images. In: Proceedings of the 2005 IEEE International Conference on Image Processing (ICIP 2005), vol. 3, pp. 65–68 (2005)
29. Popescu, A.C., Farid, H.: Exposing digital forgeries in color filter array interpolated images. IEEE Transactions on Signal Processing 53(10), 3948–3959 (2005)
30. Johnson, M.K., Farid, H.: Exposing digital forgeries through chromatic aberration. In: MM&Sec 2006, Proceedings of the Multimedia and Security Workshop 2006, September 26-27, pp. 48–55. ACM Press, New York (2006)
31. Mondaini, N., Caldelli, R., Piva, A., Barni, M., Cappellini, V.: Detection of malevolent changes in digital video for forensic applications. In: Delp, E.J., Wong, P.W. (eds.) Proceedings of SPIE: Security and Watermarking of Multimedia Content IX, vol. 6505, p. 65050T (2007)
32. Popescu, A.C., Farid, H.: Exposing digital forgeries by detecting duplicated image regions. Technical Report TR2004-515, Department of Computer Science, Dartmouth College, Hanover, NH, USA (2004)
33. Popescu, A.C., Farid, H.: Exposing digital forgeries by detecting traces of resampling. IEEE Transactions on Signal Processing 53(2), 758–767 (2005)
34. Kirchner, M.: Fast and reliable resampling detection by spectral analysis of fixed linear predictor residue. In: MM&Sec 2008, Proceedings of the Multimedia and Security Workshop 2008, September 22-23, 2008, pp. 11–20. ACM Press, New York (2008)
35. Wang, W., Farid, H.: Exposing digital forgeries in video by detecting duplication. In: MM&Sec 2007, Proceedings of the Multimedia and Security Workshop 2007, Dallas, TX, USA, September 20-21, pp. 35–42 (2007)
36. Johnson, M.K., Farid, H.: Exposing digital forgeries in complex lighting environments. IEEE Transactions on Information Forensics and Security 2(3), 450–461 (2007)
37. Böhme, R.: An epistemological approach to steganography. In: accepted for Information Hiding 2009, Darmstadt, Germany, June 7–10. LNCS (to appear, 2009)
38. Joyce, R., Gupta, G.: Identity authentication based on keystroke latencies. Communications of the ACM 33, 168–176 (1990)
39. Kerckhoffs, A.: La cryptographie militaire. Journal des sciences militaires IX, 5–38, 161–191 (1883)
40. Harris, R.: Arriving at an anti-forensics consensus: Examining how to define and control the anti-forensics problem. Digital Investigation 3(suppl. 1), 44–49 (2006)
41. Lukáš, J., Fridrich, J., Goljan, M.: Digital camera identification from sensor noise. IEEE Transactions on Information Forensics and Security 1(2), 205–214 (2006)
42. Gloe, T., Kirchner, M., Winkler, A., Böhme, R.: Can we trust digital image forensics? In: MULTIMEDIA 2007: Proceedings of the 15th international conference on Multimedia, September 24–29, 2007, pp. 78–86. ACM Press, New York (2007)
43. Kirchner, M., Böhme, R.: Synthesis of color filter array pattern in digital images. In: Delp, E.J., Dittmann, J., Memon, N.D., Wong, P.W. (eds.) Proceedings of SPIE-IS&T Electronic Imaging: Media Forensics and Security XI, vol. 7254, p. 725421 (2009)

Using Sensor Noise to Identify Low Resolution Compressed Videos from YouTube

Wiger van Houten and Zeno Geradts

Netherlands Forensic Institute
Laan van Ypenburg 6, 2497 GB
The Hague, The Netherlands
{w.van.houten,z.geradts}@nfi.minjus.nl

Abstract. The Photo Response Non-Uniformity acts as a digital fingerprint that can be used to identify image sensors. This characteristic has been used in previous research to identify scanners, digital photo cameras and digital video cameras. In this paper we use a wavelet filter from Lukáš et al [1] to extract the PRNU patterns from multiply compressed low resolution video files originating from webcameras after they have been uploaded to YouTube. The video files were recorded with various resolutions, and the resulting video files were encoded with different codecs. Depending on video characteristics (e.g. codec quality settings, recording resolution), it is possible to correctly identify cameras based on these videos.

Keywords: Photo Response Non Uniformity, Video Camera Identification, Pattern Noise, Digital Forensics, YouTube, Low resolution, MSN Messenger, Windows Live Messenger.

1 Introduction

CCD and CMOS image sensors are being integrated in a wide range of electronics, such as notebooks and mobile phones, apart from the 'traditional' electronics as video- or photo-cameras. When these videos or photographs are used in a legal context, an interesting question to be answered remains: who recorded or photographed the depicted scene? To answer this question, it may be of great help to establish which specific camera was used to record the scene. Although modern cameras often write metadata to the file, there is currently no standard framework for this kind of information for video files.

Traditionally, defective pixels on the image sensor could be used to establish the image origin. The location of these defective pixels act as a fingerprint, as these defective pixels are present in all images the sensor produces [2]. A problem with this method occurs when no defective pixels are present, or when the acquisition device internally corrects these defective pixels during post-processing. Instead of looking at the location of the defective pixels, we now look at the individual pixels that may report slightly lower or higher values compared to their

Z.J.M.H. Geradts, K.Y. Franke, and C.J. Veenman (Eds.): IWCF 2009, LNCS 5718, pp. 104–115, 2009.

neighbours, when these pixels are illuminated uniformly. These small deviations form a device signature, and this pattern is used for device identification.

The origins of this pattern noise suggest that each sensor has its own unique pattern. By extracting and comparing these patterns, device identification is possible in the same way as fingerprint identification, i.e. by comparing the pattern from a questioned image with the reference patterns from multiple cameras. When two patterns show a high degree of similarity, it is an indication that both patterns have the same origin. We conjecture that it is advantageous to build a database of patterns obtained e.g. from videos depicting child pornography; in this way the origin of different videos can be linked together.

The Photo Response Non-Uniformity (PRNU) has been successfully used in digital photo-camera identification [3,4,5], scanner identification [6,7] and also in camcorder identification [8]. Digicams often also allow to record video, but the patterns may be weak, due to subsampling or binning of the pixels in the output video.

This paper is organised as follows. In Sect. 2 the origins of the PRNU are briefly mentioned. In Sect. 3 the algorithm from [1] used to extract the PRNU pattern from images and videos is explained. In Sect. 4 we apply the method to low resolution videos from webcams that are subsequently uploaded to YouTube. These videos were initially compressed by XVID or WMV before they were uploaded. Finally, in Sect. 5 we conclude this paper.

2 Sensor Noise Sources

During the recording of the scene on the CMOS Active Pixel Sensors (APS) or CCD image sensor, various noise sources degrade the image. A distinction can be made between temporal and spatial noise. The temporal noise can be reduced by averaging multiple frames. For a comprehensive overview of noise sources in CCD and CMOS digital (video) cameras, see [12,13] and the references therein.

The main contribution to the temporal noise comes from the (photonic) shot noise and in lesser extent to the (thermal) dark current shot noise. The photonic shot noise is inherent to the nature of light, and is essentially a quantum mechanical effect due to the discrete electromagnetic field energy [14]. The dark current shot noise follows a similar distribution, and originates from the thermal generation of charge carriers in the silicon substrate of the image sensor. As the camera is not able to differentiate the signal charge from the spurious electrons generated, these unwanted electrons are added to the output and represent a noise source. In CMOS active pixel sensors additional sources are present due the various transistors integrated on each pixel [15,16,17].

Some of the variations due to dark current are somewhat systematic: the spatial pattern of these variations remains constant. Crystal defects, impurities and dislocations present in the silicon lattice may contribute to the size of this Fixed Pattern Noise (FPN), as well as the detector size, non-uniform potential wells and varying oxide thickness.

A source somewhat similar in characteristics to FPN is PRNU, the variation in pixel response when the sensor *is* illuminated. This variation comes e.g. from non-uniform sizes of the active area where photons can be absorbed, a linear effect. Another possibility is the presence of non-uniform potential wells giving a varying spectral response. If the potential well is locally shallow, long wavelength photons may not be absorbed. In principle it is possible to remove the PRNU, or even add the pattern of a different camera [18]. It is also possible to reduce the PRNU inside the camera by a form of non-uniformity correction [19].

FPN together with PRNU form the pattern noise and is always present, though in varying amount due to varying illumination between successive frames and the multiplicative nature of the latter noise source.

There are also noise sources that do not find their origin on the image sensor but are added further down the pipeline, i.e. when the digital signal is processed. For example, the quantisation noise when the potential change detected for each pixel is digitised in the analogue-to-digital converter, or when the individual pixels are interpolated to give each pixel its three common RGB values (demosaicing). This interpolation gives small but detectable offsets, and can be seen as a noise source [20,21]. Cameras of the same type or model may for this reason have slightly similar estimated noise patterns. Hence, it is advised to include a large number of cameras of the same type in actual casework. In [29] a large scale test is performed using 6896 cameras from 150 different models with images from Flickr. It is found that the error rates do not increase when multiple cameras of the same model are used. However, with the low quality cameras we intend to use, this is not the case: common compression artefacts can be found in the estimated pattern. These characteristics can in fact be used in the process of device classification [20,30,31]. Therefore, a higher than expected correlation can be found between cameras of the same type. Large scale testing with videos is more problematic, as no online databases exist with the required information, such as the EXIF metadata used in [29] to establish the ground truth.

3   Extracting the PRNU Pattern

3.1   Introduction

The goal of a denoising filter is to suppress or remove noise, without substantially affecting the (small) image details. The pattern noise in digital images can be seen as a non-periodic signal with sharp discontinuities, as the PRNU is a per-pixel effect. To extract the pattern noise from the image, a wavelet transform is used:

$$\mathcal{W}f(\tau, s) = \int_{-\infty}^{+\infty} f(t)\frac{1}{\sqrt{s}}\psi^*\left(\frac{t-\tau}{s}\right)dt \ , \tag{1}$$

where $\mathcal{W}$ denotes the wavelet transform, and ψ the mother wavelet. By scaling and translating this mother wavelet different 'window' functions are obtained, the 'daughter wavelets'. By scaling the mother wavelet the wavelet is dilated or compressed (the 'window' function is resized), and by translating the wavelet the location of the window is changed. The coarsest scale (large s, a 'window' with large support) detects low frequencies, the approximation details. On the contrary, a fine scale is sensitive to high frequencies, the detail coefficients, as can be seen from the formula. Each scale represents a different subband, as can be seen in Fig. 1a. Due to the separate time- and scale parameters, the wavelet functions are localised in space and in frequency. When a wavelet coefficient is large, a lot of signal energy is located at that point, which may indicate important image features such as edges. On the other hand, when a wavelet coefficient is small, the signal does not strongly correlate with the wavelet which means a low amount of signal energy is present and indicates smooth regions or noise. We employ the denoising filter as presented by Lukáš *et al* [1], which in turn is based on the work presented in [9], in which an algorithm used for image compression is used for image denoising. The presented algorithm was implemented using the free WaveLab package [24] in Matlab, and has been integrated in the open source NFI PRNUCompare program that has been made freely available [25].

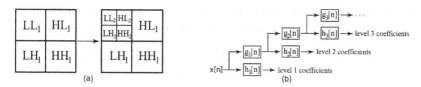

(a)   (b)

Fig. 1. (a) Subbands of a two dimensional wavelet transform. After the approximation and detail coefficients are calculated, the approximation details (LL_1) are split up in high- and low frequency subbands again. The end result is an image with the same size as the input image, composed of nested submatrices each representing a different detail level. (b) Iterated filterbank. The output of the lowpass filter g is the input of the next stage.

3.2   Algorithm

First, the video is split up in individual frames using FFmpeg [26]. The image is assumed to be distorted with zero-mean White Gaussian Noise (WGN) N(0, σ_0^2) in the spatial domain with variance σ_0^2, and hence this noise is also WGN after transforming it into the wavelet domain.

1. The fourth level wavelet decomposition using the Daubechies wavelet is obtained by letting a cascade of filters (that form a Quadrature Mirror Filter) work on the image data, decomposing the image into an orthonormal basis (transform coding). We have seen that the Coiflet wavelet may have a slightly better performance. The level-1 approximation coefficients are obtained by filtering the image data through a low-pass filter g, while the level-1 detail coefficients are obtained by filtering the image data through a high-pass filter h. By filtering the level-1 approximation coefficients (LL$_1$ subband) with the same set of filters g and h, the level-2 approximation and detail coefficients are produced by iteration, as represented in Fig. 1b (see e.g. Ch. 5 of [27]). Each resolution and orientation has its own subband, with HL$_1$ representing the finest details at scale 1 where the high pass filter was applied in the

• • •

Fig. 2. Left is the low resolution residual. The images are obtained by applying the inverse wavelet transform to the wavelet representation of different scales. Moving to the right more detail is added (lower scales) until the final image is obtained.

horizontal direction and the lowpass filter in the vertical direction. LL_4 represents the low resolution residual.

This wavelet decomposition into different detail and approximation levels allows the image to be represented as a superposition of coarse and small details, as schematically represented in Fig. 2.

2. For all pixels in each subband the local variance is estimated for each coefficient with a variable size square neighbourhood N with size $W \in (3, 5, 7, 9)$.

$$\hat{\sigma}_W^2(i, j) = \max\left(0, \frac{1}{W^2} \sum_{(i,j) \in N} LH_s^2(i, j) - \sigma_0^2\right) , \qquad (2)$$

with (i, j) representing the pixel location in each subband. This estimates the local signal variance in each subband, and the minimum variance of each pixel for these varying size neighbourhoods is taken as the final estimate:

$$\hat{\sigma}^2(i, j) = \min(\sigma_{w \in W}^2(i, j)) . \qquad (3)$$

3. The wavelet coefficients in the detail subbands can be represented by a generalised Gaussian with zero mean [28]. This σ_0^2 parameter controls how strong the noise suppression will be. When we estimate the reference pattern as well as when we estimate the pattern noise from the (questioned) natural video (a video that contains some sort of texture, like in a normal scene), we need to set this parameter (denoted σ_{ref} and σ_{nat} respectively).

The actual denoising step takes place in the wavelet domain by attenuating the low energy coefficients as they are likely to represent noise. This is done in all detail subbands with the Wiener filter:

$$LH_s(i, j) = LH_s(i, j) \frac{\hat{\sigma}^2(i, j)}{\hat{\sigma}^2(i, j) + \sigma_0^2} . \qquad (4)$$

4. The above steps are repeated for all levels and colour channels. By applying the inverse discrete wavelet transform to the denoised coefficients, the denoised image is obtained. By subtracting this denoised image from the original input image, the estimated PRNU pattern is obtained. As a final step this pattern is zero-meaned such that the row and column averages are zero by subtracting the column averages from each pixel and subsequently subtracting the row averages from each pixel. This is done to remove artefacts from e.g. colour interpolation, as suggested in [4].

3.3   Obtaining the Sensor Noise Patterns and Detecting the Origin

To determine whether a specific questioned video V_q originates from a certain camera C, we first extract the individual frames I_{q_i} ($i = 1 \ldots N$, with N the amount of frames in V_q) from the video, and subtract the denoised image I_{d_i} from each individual frame:

$$p_{q_i} = I_{q_i} - I_{d_i} \text{ , with } I_{d_i} = \mathcal{F}(I_{q_i}) \text{ ,} \tag{5}$$

and $\mathcal{F}$ the filter as described above. To obtain the PRNU pattern, we use the maximum likelihood estimator, as derived in [5]:

$$p_q = \frac{\sum_{i=1}^{N} p_{q_i} I_{q_i}}{\sum_{i=1}^{N} I_{q_i}^2} \text{ ,} \tag{6}$$

with element-wise multiplication implied. In a similar manner the reference patterns p_{r_j} from different cameras with a known origin are calculated for different cameras. For this task, it is preferred to use a flatfield video V_f from which individual flatfield images I_{f_i} can be extracted that have no scene content and approximately uniform illumination. This is done for multiple cameras, each with its own reference pattern p_{r_j}. After all the reference patterns are obtained, the final step is to measure the degree of similarity between the questioned pattern and the reference patterns. We use the total correlation (summed over all colour channels) as the similarity measure in order to find out whether a certain pattern p_q originates from a certain camera C. When the correlation of p_q is highest for a certain p_{r_j}, we decide the video was most likely acquired using camera j.

4   Application to YouTube Videos

YouTube is a website where users can view and share (upload) video content. Videos encoded with the most popular encoders are accepted as upload, after which the uploaded video is converted to a format that may be viewed in a Flash player. The videos can be downloaded by using services as keepvid.com.

To assess the performance of the algorithm for videos that are uploaded to YouTube, we uploaded multiple (natural) videos encoded with different settings and from different web cameras to YouTube. The flatfield video was obtained by recording (without any form of additional compression on top of the possibly compressed output from the cameras) a flat piece of paper under various angles in order to vary the Discrete Cosine Transform (DCT) coefficients in the compression blocks for the duration of approximately 30 seconds. Natural video (also approximately 30 seconds) was obtained by recording the surroundings of the office in which scenes with a high amount of details alternated smooth scenes, both with dark and well-illuminated scenes. Static shots alternated shots with fast movements, and saturation occurred frequently. When the resolution of the uploaded (natural) content exceeds the maximum resolution that can be obtained from YouTube, YouTube resizes the input video. In this case, it is also necessary to resize the frames from the reference videos.

The patterns obtained from each natural video are compared with the reference patterns from all other cameras of the same type. Although this is a limited test, we avoid the problems of incompatible resolutions. A future test without this assumption is planned, but the processing time will increase significantly.

As explained above, the σ-parameters control the amount of noise that is extracted from each frame. To see which settings perform best, we calculate the reference patterns as well as the natural patterns (the patterns obtained from the natural video) for multiple values: $\sigma_{\text{nat}} = 0.5 + n$, $(n = 0 \ldots 8)$, $\sigma_{\text{ref}} = 0.5 + r$, $(r = 0 \ldots 8)$. By calculating the correlation between all these possible pairs we can find the optimum parameters. In actual casework this is not possible, as the questioned video has an unknown origin. We only report the correlation values of the matching (the natural video and reference material have the same origin) and the maximum correlation value of the mismatching pairs (the maximum correlation between the pattern from the natural video and the patterns from all other unrelated cameras), ρ_m and ρ_{mm} respectively. Hence, when ρ_m is larger than ρ_{mm} the correct camera was indeed identified (the highest correlation is presented in bold font). We tested several web cameras, but for reasons of brevity we only report extensively upon one type, namely the Logitech Communicate STX, and only briefly comment on the other cameras and tests.

Remark

At the time these calculations were done, the maximum available resolution that could be viewed was 480x360 (aspect ratio 4:3). Recently the maximum resolution that can be viewed (and downloaded) was increased for higher resolution videos.

4.1   Logitech Quickcam STX

We recorded for each of the 8 Logitech Quickcam STX cameras a 30 second sample with natural content in the native resolution of 640x480 with the XVID codec with quality setting 4, as well as a 30 second flatfield sample in the same resolution in RAW. As we want to compare the noise pattern extracted from these natural videos with the noise patterns extracted from the flatfield videos, we also need to resize the flatfield videos, as this video can only be downloaded in 480x360 resolution. It is recommended to record the video in native resolution, and subsequently resize the individual frames from the flatfield video (e.g. with a bilinear resize) to match the resolution of the natural video. Regardless of the σ-parameter settings this resulted in a 100% correct identification rate as can be seen in Table 1.

We repeated the experiment with the same cameras and only changed the recording resolution to 320x240. Recording in a lower than native resolution means in this case that the pixels in the output video are binned (in this case 4 pixels are averaged to give the output of 1 pixel) which results in a strong

Table 1. Logitech Communicate STX. Frames from the uncompressed flatfield videos are resized with a bilinear interpolation to match the size of the natural video downloaded from YouTube (480x360), after which the reference patterns are calculated with parameters $\sigma_{nat} = 6.5$, $\sigma_{flat} = 7.5$.

	cam1	cam2	cam3	cam4	cam5	cam6	cam7	cam8
ρ_m	**0.1334**	**0.2301**	**0.1279**	**0.1818**	**0.1596**	**0.1622**	**0.1515**	**0.2099**
ρ_{mm}	0.0374	0.0512	-0.0009	0.0342	0.0355	0.0290	0.0118	0.0421

Table 2. Logitech Communicate STX. As the natural video downloaded from YouTube is not resized (320x240), the reference patterns are estimated directly from the uncompressed video with parameters $\sigma_{nat} = 3.5$, $\sigma_{flat} = 5.5$.

	cam1	cam2	cam3	cam4	cam5	cam6	cam7	cam8
ρ_m	**0.1044**	**0.0936**	**0.1090**	0.0153	**0.0304**	**0.1044**	**0.0984**	0.0334
ρ_{mm}	0.0280	0.0616	0.0803	**0.0407**	0.0245	0.0526	0.0652	**0.0648**

attenuation of the PRNU, as the PRNU is a per-pixel effect. If one general set of parameters is chosen, a maximum of 6 cameras were correctly identified, as can be seen in Table 2.

Codec Variations. For one camera we recorded video in the native resolution of 640x480, as well as the lower resolution 320x240 for two different codecs (WMV and XVID) and different codec-settings. In order to let the video content be the same for all videos, we first recorded the video in RAW at both resolutions, and subsequently encoded it with different codec settings in VirtualDub [32]. For the XVID codec we used quality settings $Q = 4 \cdot n$, with $n = 1 \ldots 8$, while for the WMV9 codec we used quality settings $Q\prime = 10 \cdot n$, $n = 5 \ldots 9$. Note that in the case of XVID higher Q values represents higher compression, while in the case of the WMV9 codec a higher setting means higher quality. The videos were uploaded to YouTube, and subsequently downloaded after which the sensor pattern noise was extracted again.

For the native resolution, the correct identification rate is 100%, regardless of the codec and setting used. When the recording resolution is set to 320x240 we see that the correct identification rate is lowered (Tables 3 and 4).

Video Extracted from Windows Live Messenger Stream. Windows Live Messenger, formerly known as MSN Messenger, is a popular instant messaging client, which provides webcam support. Through the use of external programs it is possible to record the video stream sent during a webcam session, often simply by capturing the screen. It is also possible to directly record the data from the stream, as is done with MSN Webcam Recorder [33].

As a final test with this webcam, we set up a webcam session between two computers with Windows Live Messenger, with one computer capturing a web-

Table 3. Logitech Communicate STX. Video recorded in 320x240 with the XVID codec, variable quality Q (filesize indicated in kilobytes (kB)).$\sigma_{nat} = 5.5$, $\sigma_{flat} = 4.5$

Table 4. Logitech Communicate STX. Video recorded in 320x240 with the WMV9 codec, variable quality Q/ (filesize indicated in kilobytes (kB)). $\sigma_{nat} = 6.5$, $\sigma_{flat} = 7.5$

Q	kB	kbit/s	bpp	ρ_m	ρ_{mm}
4	2206	535	0.859	**0.1173**	0.0497
8	1238	300	0.429	0.0795	**0.0852**
12	949	230	0.311	**0.1115**	0.0541
16	813	197	0.255	**0.0811**	0.0608
20	750	182	0.221	**0.1474**	0.0472
24	703	171	0.205	**0.0935**	0.0684
28	675	164	0.190	**0.1259**	0.0531
32	660	160	0.184	**0.1026**	0.0381

Q/	kB	kbit/s	bpp	ρ_m	ρ_{mm}
90	3023	734	0.859	**0.0939**	0.0811
80	1717	417	0.428	**0.1107**	0.0894
70	1229	298	0.310	**0.1483**	0.0823
60	953	231	0.254	**0.1001**	0.0800
50	815	198	0.221	**0.0663**	0.0481
40	700	170	0.204	0.0592	**0.0740**

cam stream of approximately two minutes sent out by the other computer. The stream was sent out as a WMV9 video at a resolution of 320x240 (selected as 'large' in the host client). After the data was recorded with the aforementioned program, it was encoded with the XVID codec with a bitrate of 200 kbps, which resulted in 1705-1815 frames (0.17-0.18 bpp). Finally, the resulting video was uploaded to YouTube, where a third layer of compression was added. It has to be stressed that in practice with low bandwidth systems the framerate may be reduced significantly. We again see the source camera is correctly identified, in Table 5, but that the correlations for matching and mismatching pairs are close.

Table 5. Logitech Communicate STX. Video (320x240) recorded from webcam stream from Windows Live Messenger (WMV9) and subsequently encoded with the XVID codec. $\sigma_{nat} = 4.5$, $\sigma_{ref} = 2.5$

	cam1	cam2	cam3	cam4	cam5	cam6	cam7
ρ_m	**0.1029**	**0.1361**	**0.0792**	**0.1060**	**0.1010**	**0.0770**	**0.0616**
ρ_{mm}	0.0421	0.0383	0.0476	0.0129	0.0459	0.0288	0.0505

4.2   Other Cameras

For 6 Creative Live! Cam Video IM webcams we recorded a 30 second sample in 352x288 (11:9), while the native resolution is 640x480 (4:3). These videos were encoded using the WMV9 codec, with quality setting 70 and uploaded to YouTube. This resulted in videos with a bitrate between 180 and 230 kbit/s (0.13-0.16 bpp). Only 5 out of 6 cameras were correctly identified. Next, we recorded a video with resolution of 800x600 (4:3) and encoded it with quality setting 60 in WMV9. After recording the flatfield videos in the native resolution, we resized the individual frames from the flatfield videos. This resulted in a

100% correct identification rate, although the difference between matching and mismatching correlations is small.

For each of the ten Vodafone 710 mobile phones, we recorded a 30 second sample in the native resolution of 176x144. This phone stores the videos in the 3GP format. This is, like the AVI file format, a container format in which H.263 or H.264 can be stored. This phone uses the H.263 format optimised for low-bandwidth systems. The natural video had a bitrate between 120 and 130 kbit/s (0.36-0.39 bpp). After uploading the natural videos, YouTube automatically changed the aspect ratio, from 11:9 to 4:3. The correct identification rate is only 50%, possibly due to the codec used to initially encode the video: the H.263 codec uses a form of vector quantisation, which is different from the DCT transform used in WMV and XVID.

5   Conclusion

We have seen that it is possible to identify low resolution webcams based on the extraction of sensor noise from the videos it produces, even after these videos were re-compressed by YouTube. Depending on various video characteristics identification is possible in a wide range of circumstances.

As there are a lot of parameters it is not possible to give a general framework to which a video should comply in order for a correct identification to occur. In general, by setting the parameter for extracting the PRNU pattern from natural or flatfield videos between 4 and 6, satisfactory results are obtained. It has to be kept in mind that the noise patterns estimated from cameras of the same make and model may contain similarities. Especially as the output of webcams or mobile phones may actually be a compressed JPEG stream in which the DCT compression blocks can be clearly distinguished, this may introduce higher than expected correlations based on the sensor noise alone. For this reason it is advised to use a large amount of cameras of the same brand and model in actual casework. In this way we can see whether similarities between noise patterns comes from the actual sensor noise, or whether it comes from post-processing steps. A larger scale test will be done in the near future, so we can hopefully give figures for the False Rejection and False Acceptance Ratio.

In general, we have two opposing predictions for the future with respect to the source video camera identification. On one hand the sensor noise is likely to decrease when manufacturing standards increase. On the other hand, with increasing resolution (smaller pixels) the relative PRNU size may increase as well. Also, the storage capacity of these devices (in mobile phones, but also the maximum allowed video size that can be uploaded to e.g. YouTube) increases. Furthermore, the bandwidth of (mobile) networks increases, allowing faster transfers of these files. Finally, due to the increasing processing power in these electronic devices, more advanced compression schemes become available, possibly retaining more of the pattern noise after compression.

Acknowledgement

This work was partially supported by the European Union sponsored FIDIS Network of Excellence. We thank the reviewers for their time and effort.

References

1. Lukáš, J., Fridrich, J., Goljan, M.: Digital Camera Identification from Sensor Pattern Noise. IEEE Trans. on Information Forensics and Security 1, 205–214 (2006)
2. Geradts, Z., Bijhold, J., Kieft, M., Kurosawa, K., Kuroki, K., Saitoh, N.: Methods for Identification of Images Acquired with Digital Cameras. In: Proc. SPIE, vol. 4232 (2001)
3. Alles, E.J., Geradts, Z.J.M.H., Veenman, C.J.: Source Camera Identification for Low Resolution Heavily Compressed Images. In: Int. Conf on Computational Sciences and Its Applications, 2008. ICCSA 2008, pp. 557–567 (2008)
4. Chen, M., Fridrich, J., Goljan, M.: Digital Imaging Sensor Identification (Further Study). In: Proceedings of the SPIE, Security, Steganography, and Watermarking of Multimedia Contents IX, vol. 6505, p. 65050P (2007)
5. Chen, M., Fridrich, J., Goljan, M., Lukáš, J.: Determining Image Origin and Integrity Using Sensor Noise. IEEE Trans. on Information Forensics and Security 3(1), 74–90 (2008)
6. Khanna, N., Mikkilineni, A.K., Chiu, G.T.C., Allebach, J.P., Delp, E.J.: Scanner Identification Using Sensor Pattern Noise. In: Proc. SPIE, Security, Steganography, and Watermarking of Multimedia Contents IX, vol. 6505, p. 65051K (2007)
7. Gou, H., Swaminathan, A., Wu, M.: Robust Scanner Identification Based on Noise Features. In: Proc. SPIE, Security, Steganography, and Watermarking of Multimedia Contents IX, vol. 6505, p. 65050S (2007)
8. Chen, M., Fridrich, J., Goljan, M., Lukáš, J.: Source Digital Camcorder Identification using Sensor Photo Response Non-Uniformity. In: Proc. SPIE, vol. 6505 (2007)
9. Mihçak, M.K., Kozintsev, I., Ramchandran, K.: Spatially Adaptive Statistical Modeling of Wavelet Image Coefficients and its Application to Denoising. In: Proc. IEEE Int. Conf. Acoust. Speech, and Signal Proc., pp. 3253–3256 (1999)
10. Johnson, M.K., Farid, H.: Exposing digital forgeries through chromatic aberration. MM&Sec 2006. In: Proceedings of the Multimedia and Security Workshop 2006, Geneva, Switzerland, September 26-27, pp. 48–55 (2006)
11. Van, L.T., Emmanuel, S., Kankanhalli, M.S.: Identifying Source Cell Phone Using Chromatic Aberration. In: Proceedings of the 2007 IEEE International Conference on Multimedia and EXPO (ICME 2007), pp. 883–886 (2007)
12. Holst, G.C., Lomheim, T.S.: CMOS / CCD Sensors and Camera Systems. JCD Publishing and SPIE Press (2007)
13. Irie, K., McKinnon, A.E., Unsworth, K., Woodhead, I.M.: A Model for Measurement of Noise in CCD Digital-Video Cameras. Measurement Science and Technology 19, 45207 (2008)
14. Loudon, R.: Quantum Theory of Light. Oxford University Press, Oxford (2001)
15. Tian, H., Fowler, B.A., El Gamal, A.: Analysis of Temporal Noise in CMOS APS. In: Proc. SPIE Vol. 3649, Sensors, Cameras, and Systems for Scientific/Industrial Applications, pp. 177–185 (1999)

16. Salama, K., El Gamal, A.: Analysis of Active Pixel Sensor Readout Circuit. IEEE Trans. on Circuits and Systems I, 941–945 (2003)
17. El Gamal, A., Fowler, B., Min, H., Liu, X.: Modeling and Estimation of FPN Components in CMOS Image Sensors. In: Morley, M.M. (ed.) Proc. SPIE, Solid State Sensor Arrays: Development and Applications II, vol. 3301, pp. 168–177 (1998)
18. Gloe, T., Kirchner, M., Winkler, A., Böhme, R.: Can We Trust Digital Image Forensics? In: Proc. of 15th International Conference on Multimedia, pp. 79–86 (2007)
19. Ferrero, A., Campos, J., Pons, A.: Correction of Photoresponse Nonuniformity for Matrix Detectors Based on Prior Compensation for Their Nonlinear Behavior. Applied Optics 45(11) (2006)
20. Bayram, S., Sencar, H.T., Memon, N., Avcıbaş, İ.: Source Camera Identification Based on CFA Interpolation. In: ICIP 2005, vol. 3, pp. 69–72 (2005)
21. Long, Y., Huang, Y.: Image Based Source Camera Identification using Demosaicking. In: IEEE 8th Workshop on Multimedia Signal Processing, October 3-6, pp. 419–424 (2006)
22. Donoho, D.L., Johnstone, I.M.: Ideal Spatial Adaptation by Wavelet Shrinkage. Biometrika 81, 425–455 (1994)
23. Kaiser, G.: A Friendly Guide to Wavelets. Birkhauser Boston Inc, Basel (1994)
24. Donoho, D., Maleki, A., Shahram, M.: WaveLab 850, http://www-stat.stanford.edu/~wavelab/
25. van der Mark, M., van Houten, W., Geradts, Z.: NFI PRNUCompare, http://prnucompare.sourceforge.net
26. FFmpeg, free program to convert audio and video, http://www.ffmpeg.org
27. Mallat, S.: A Wavelet Tour of Signal Processing, 2nd edn. Academic Press (1999)
28. Chang, S.G., Yu, B., Vetterli, M.: Adaptive Wavelet Thresholding for Image Denoising and Compression. IEEE Trans. on Image Processing 9, 1532–1546 (2000)
29. Goljan, M., Fridrich, J., Filler, T.: Large Scale Test of Sensor Fingerprint Camera Identification. In: Proc. SPIE, Electronic Imaging, Security and Forensics of Multimedia Contents XI, San Jose, CA, January 18-22 (2009)
30. Bayram, S., Sencar, H.T., Memon, N.: Improvements on Source Camera-Model Identification Based on CFA Interpolation. In: Proc. WG 11.9 Intl. Conf. on Digital Forensics (2006)
31. Çeliktutan, O., Avcıbaş, İ., Sankur, B.: Blind Identification of Cellular Phone Cameras. In: Proceedings of the SPIE, Security, Steganography, and Watermarking of Multimedia Contents IX, vol. 6505, p. 65051H (2007)
32. Virtualdub: A free AVI/MPEG-1 processing utility, http://www.virtualdub.org
33. MSN Webcam Recorder, http://ml20rc.msnfanatic.com/
34. Goljan, M., Fridrich, J.: Camera Identification from Cropped and Scaled Images. In: Proc. SPIE, Electronic Imaging, Forensics, Security, Steganography, and Watermarking of Multimedia Contents X, January 26-31, pp. OE-1–OE-13. San Jose (2008)
35. Çeliktutan, O., Sankur, B., Avcıbaş, İ.: Blind Identification of Source Cell-phone Model. IEEE Trans. on Information Forensics and Security 3, 553–566 (2008)

Using the ENF Criterion for Determining the Time of Recording of Short Digital Audio Recordings

Maarten Huijbregtse and Zeno Geradts

Netherlands Forensic Institute, Departement Digital Evidence and Biometrics,
Laan van Ypenburg 6, 2497 GB DEN HAAG Netherlands
Z.geradts@nfi.minjus.nl

Abstract. The Electric Network Frequency (ENF) Criterion is a recently developed forensic technique for determining the time of recording of digital audio recordings, by matching the ENF pattern from a questioned recording with an ENF pattern database. In this paper we discuss its inherent limitations in the case of short – i.e., less than 10 minutes in duration – digital audio recordings. We also present a matching procedure based on the correlation coefficient, as a more robust alternative to squared error matching.

Keywords: ENF, authentication, integrity.

1 Introduction

Electric networks operate at their own specific frequency: the Electric Network Frequency (ENF)[1]. However, due to unbalances in production and consumption of electrical energy, the ENF is known to fluctuate slightly over time rather than being stuck to its exact set point (figure 1). The fluctuation pattern is the same throughout the entire network [2] [3].

Digital recording equipment – both mains and battery powered – can pick up the ENF[2], which ends up as an extra frequency component in the recorded audio file [2] [3] [4]. By band pass filtering the audio signal, the ENF can be isolated and its pattern can be retrieved. Under the assumption that the ENF fluctuations are random, this effectively puts a time-stamp on the audio recording: the ENF pattern is unique for the time at which the recording was made.

The ENF Criterion

One of the challenges in authenticating digital audio evidence is to gain insight into its time of recording [5]. A technique known as the *ENF criterion* [2] uses the

---

[1] The main part of continental Europe is served by one large electric network, controlled by the UCTE [1]. Its ENF is set at 50 Hz.

[2] Claims are that recording equipment's microphones are sensitive to the power socket signal (when mains powered) and the electromagnetic fields emanating from nearby power lines (when battery powered). A thorough investigation of recording equipment for which these claims hold is, however, lacking.

Z.J.M.H. Geradts, K.Y. Franke, and C.J. Veenman (Eds.): IWCF 2009, LNCS 5718, pp. 116–124, 2009.
© Springer-Verlag Berlin Heidelberg 2009

aforementioned ENF fluctuation to achieve this[3]. By comparing the recorded ENF pattern to a database ENF pattern from the same electric network, it is possible to:

1) verify (or falsify) a questioned time of recording, or
2) determine an unknown time of recording.

A visual comparison of the recorded and database ENF patterns is often adequate for the first case, while an (automated) search routine is necessary for the latter, to locate the best match between recorded and database pattern[4].

Paper Outline

Up until now, the ENF criterion has mostly been used with long recordings – i.e., approximately one hour in duration [2] [3] [4]. However, the amount of digital audio evidence (often accompanied by video footage) of short duration – i.e., ten minutes or less – is increasingly prominent with the advent of audio and video recording capabilities in consumer products (e.g., cell phones and digital cameras).

In the second half of this paper (sections 4 and 5) we will discuss the limitations of using the ENF criterion with short recordings, and show examples of erroneous determination of the time of recording when using a minimum squared error-based matching procedure. We present a maximum correlation coefficient-based matching procedure as a more robust alternative.

In the first half (sections 2 and 3), we will describe our means of building an ENF pattern database and extracting the ENF pattern from a digital audio recording.

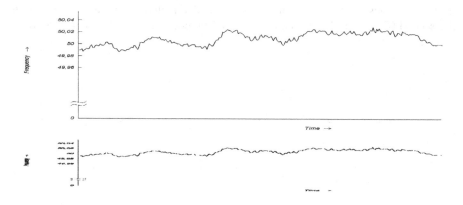

Fig. 1. ENF fluctuation over time

---

[3] See [2] for other applications of the ENF criterion.
[4] In this paper, we focus on using the ENF criterion for determining an unknown time of recording.

2 ENF Pattern Database

Since the ENF is the frequency at which voltage levels in an electric network oscillate, it is possible to obtain the ENF pattern by analyzing the voltage level signal, e.g., from a power socket. In our setup, we fed this signal – in attenuated form – to a PC sound card that was set to a sampling frequency of 8000 Hz. The sampled signal $x[n]$, with the index n starting at 1, can be modeled as:

$$x[n] = k \cdot V\big((n-1)T_s + t_1\big) \tag{1}$$

where $V(t)$ denotes the voltage level at time z^t, k is a factor representing the attenuation, T_s the sampling period (= $1.25 \cdot 10^{-4}$ s) and t_1 the time of the first sample.

We used the method of zero crossings, mentioned by Grigoras [2], for analysis of $x[n]$. The idea is to treat the signal as sinusoidal, although this is not strictly true since its frequency – the ENF – varies slightly over time. For a sinusoidal signal, the time τ between two consecutive zero crossings equals half the oscillation period, so that its inverse equals twice the oscillation frequency f :

$$f = \frac{1}{2\tau} \tag{2}$$

We determined the times of zero crossings by linear interpolation between samples $x[k]$ and $x[k+1]$ that differ in sign, and calculated the difference between two consecutive times of zero crossings to obtain values for τ. The corresponding values for f were calculated using (2) and averaged for every second of signal. This finally results in a series of frequency values – i.e., the ENF pattern – with a time resolution of 1 second (figure 2a). For visual clarity, ENF patterns are often depicted as continuous (figure 2b).

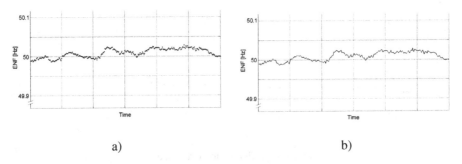

a)   b)

Fig. 2. a) ENF pattern as a series of ENF values b) Continuous ENF pattern, obtained by interpolating ENF values

3 ENF Pattern Extraction from Digital Audio Recording

We have adopted the method presented by Cooper [4] for extracting the ENF pattern from a digital audio recording. We shall cover this method briefly here, since Cooper's paper offers an excellent and comprehensive description.
The basic steps are:

- *Signal decimation* – Many digital audio recordings are recorded at high sampling frequencies – e.g., 44100 Hz. To detect the ENF, which is approximately 50 Hz, much lower sampling frequencies are allowed. The audio file is thus decimated to a sampling frequency of 300 Hz, which significantly reduces computational time.

- *Band pass filtering* – The frequencies of interest are around 50 Hz, so the decimated audio file is digitally band pass filtered from 49.5 Hz to 50.5 Hz to isolate the ENF.

- *Short Time Fourier Transform (STFT)* – In discrete time STFT analysis, a signal is divided into J partly overlapping frames (figure 3) for which, after windowing and zero-padding, the frequency spectrum is calculated via a Discrete Fourier Transform (DFT). The jump H (in samples) between frames determines the time resolution of the final ENF pattern, while the amount of overlap $M - H$ affects its smoothness. In our specific case, we have chosen $H = 300$ so that the extracted ENF pattern time resolution equals that of the database – i.e., 1 second. Each frame was windowed with a rectangular window and zero-padded by a factor of 4.

- *Peak frequency estimation* – For each frequency spectrum[5], the frequency with maximum amplitude is estimated. As it is unlikely that this 'peak frequency' coincides exactly with a DFT frequency bin, quadratic interpolation around the bin with maximum amplitude is performed. The estimated peak frequency is stored as the ENF value for the corresponding frame, so that we end up with an extracted ENF pattern of J ENF values.

4 Matching by Minimum Squared Error

Calculating the squared difference ('error') between two vectors is a common approach in determining their equivalence: the smaller the squared error, the more both vectors are alike. The squared error E for two length L vectors x and y is defined as:

$$E = \sum_{i=1}^{L} \left(x[i] - y[i] \right)^2 \tag{3}$$

When determining the time of recording using the ENF criterion, we have in general one longer vector (the database ENF pattern d) and one shorter vector (the recorded

---

[5] Actually, we used the *log power spectrum*, defined as $\log_{10} \left| X[f] \right|^2$, where $X[f]$ is the frequency spectrum.

ENF pattern r). The approach is then to calculate a vector e of squared error values, according to:

$$e[k] = \sum_{i=1}^{R} (r[i] - d[i + k - 1])^2 \qquad (4)$$

in which R is the length of the recorded ENF pattern, while the index k runs from 1 to $D - R + 1$; D being the length of the database ENF pattern. The minimum value in e determines the location of the best match between recorded and database pattern, and hence the time of recording.

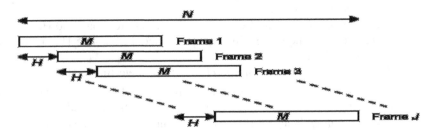

Fig. 3. Division of a signal of length N into J partly overlapping frames

Ideally, the recorded ENF pattern and its corresponding database pattern are exactly equal and the minimum error value would be zero. In practice, however, this is not the case. The reliability of the ENF criterion in determining the right time of recording is therefore limited by the occurrence of similar patterns *within the database itself* – i.e., ENF patterns with squared errors in the same range as 'typical' squared errors between recorded and corresponding database ENF pattern.

Database Analysis

In an experiment, we took roughly 1.5 years of ENF data[6] and calculated the (root mean) squared error[7] between two randomly picked, non-overlapping pieces of 600 ENF values ($\hat{=}$ 10 minutes). By repeating this one million times, we were able to picture the approximate distribution of root mean squared (rms) errors between ENF patterns of length 600 within the database (figure 4). Similar experiments were run for patterns of 60, 120, 240 and 420 ENF values.

---

[6] Collected as described in section 2, at the Netherlands Forensic Institute (The Hague, The Netherlands) from September 2005 until February 2007. ENF values were stored minute-by-minute in plain text files (i.e., 60 values per file).

[7] Following the notation of equation (3), the root mean squared error E_{rms} is defined as

$$E_{rms} = \sqrt{\frac{\sum_{i=1}^{L}(x[i] - y[i])^2}{L}}$$. Conclusions are independent of a choice for E or E_{rms}

as the dissimilarity measure.

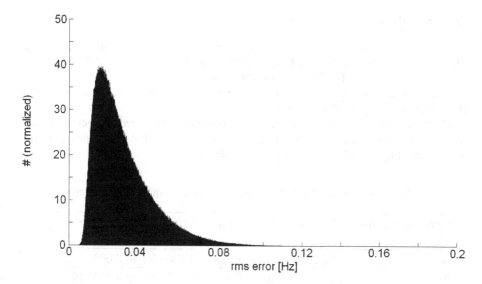

Fig. 4. Normalized histogram of squared errors between database pieces, 600 ENF values in length

The most interesting part of the histogram in figure 4 is near zero: the smallest (observed) rms error within the database ENF pattern. For length 600, we found this smallest rms error to be about 0.0040 Hz.

We thus conclude that the minimum rms error between a recorded ENF pattern of length 600 and a (large) database should be 'well below' 0.0040 Hz for a reliable determination of the time of recording.

Table 1 lists the observed smallest rms errors for all experiments. As can be expected, the error increases for longer ENF patterns: the longer a pattern, the less likely it will have a similar counterpart over its whole length within the database[8].

Table 1. Smallest observed rms errors within 1.5 years of ENF database

ENF pattern length	Smallest observed rms error [Hz]
60	0.0007
120	0.0015
240	0.0020
420	0.0035
600	0.0040

---

[8] It is therefore that the ENF criterion works well for audio recordings of long duration (as confirmed by some of our experiments not mentioned here). In this case, the rms error between recorded and corresponding database ENF pattern is almost certainly much smaller than those found within the database itself.

Test Recordings

For a second experiment, we took an "American Audio Pocket Record" portable digital audio recorder and set it up to be mains powered. We made a total of 70 recordings with durations of 60, 120, 240, 420 and 600 seconds (i.e., 14 recordings for each duration). The exact times of recording were known and the audio files – sampled at 44,1 kHz – were stored in lossless WAV format.

The ENF pattern from each recording, extracted as described in section 3, was compared to a small database consisting of two weeks of ENF data, including the period of recording. Results are summarized in table 2.

Table 2. Test recording results for minimum root-mean-squared error matching

Recording duration	Correct time estimate for...	Minimum rms error ranging from...
60 s	0 out of 14 recordings	0.0008 Hz to 0.0028 Hz
120 s	0 out of 14 recordings	0.0018 Hz to 0.0037 Hz
240 s	2 out of 14 recordings	0.0033 Hz to 0.0049 Hz
420 s	10 out of 14 recordings	0.0045 Hz to 0.0055 Hz
600 s	14 out of 14 recordings	0.0045 Hz to 0.0055 Hz

It is seen that the ENF criterion failed in correctly estimating the time of recording for 44 out of the 70 recordings. Moreover, the found minimum rms errors are all above the values mentioned in table 1. Comparison to a larger database could thus have resulted in even less satisfying results.

Matching by Maximum Correlation Coefficient

Figure 5 shows a main reason for the failure of the ENF criterion: our recorded ENF patterns have a slight offset compared to the database pattern – a phenomenon also noted by Kajstura et al [3]. In general, this cannot be known beforehand and thus the matching procedure should be robust to this type of behavior.

We propose matching based on equivalence of shape, by using the correlation coefficient. Following the notation of equation (3), the correlation coefficient ρ between two vectors is defined as:

$$\rho = \frac{\sum_{n=1}^{L}(x[n]-\bar{x})(y[n]-\bar{y})}{(L-1)\sigma_x\sigma_y} \tag{5}$$

where the horizontal bars and sigmas denote the averages and the standard deviations of the vectors respectively. ρ can run from -1 to +1; the closer the value is to +1, the more both vectors are alike in shape. When comparing a recorded ENF pattern to a database, we thus search for the *maximum correlation coefficient* between recorded and database pattern.

Database Analysis

As with minimum squared error matching, the reliability of a maximum correlation coefficient-based matching procedure will be limited by high correlations within the database itself. We have repeated the first experiment described in the preceding section, this time calculating correlation coefficients instead of rms errors. Here, we are interested in the *largest* observed values, which are listed in table 3[9].

Table 3. Largest observed correlation coefficients within 1.5 years of ENF database

ENF pattern length	Largest observed correlation coefficient
60	0. 9980
120	0.99
240	0.9870
420	0.9820
600	0.9850

Test Recordings

Matching the same 70 test recordings with the same database by a maximum correlation coefficient search, yielded the results mentioned in table 4. Correct time estimation is significantly improved, with only 3 failures out of 70. Also, from duration of 240 seconds onwards, the maximum correlation coefficients all lie above the values mentioned in table 3. This suggests that even comparisons to a larger database would have resulted in correct time estimates.

Table 4. Test recording results for maximum correlation coefficient matching

Recording duration	Correct time estimate for...	Maximum corr. coeff. ranging from...
60 s	12 out of 14 recordings	0.9758 to 0.9989
120 s	13 out of 14 recordings	0.9572 to 0.9980
240 s	14 out of 14 recordings	0.9893 to 0.9989
420 s	14 out of 14 recordings	0.9910 to 0.9992
600 s	14 out of 14 recordings	0.9945 to 0.9993

---

[9] The higher value for length 600 compared to length 420 is probably due to using 'only' one million random pairs of database pattern: the experiment for length 600 just happened to come across a better matching pattern than the one for length 420.

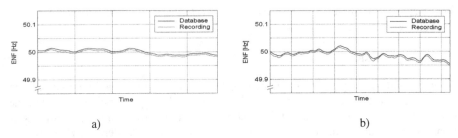

Fig. 5. Recorded ENF patterns lie consistently below the corresponding database patterna) Example of a recording 240 s in duration b) Example of a recording 420 s in duration

6 Conclusion

We have shown that the reliability of the ENF criterion is inherently limited by similarities within the ENF pattern database to which the recording is compared. The possible presence of a frequency offset further increases the danger of erroneous determination of the time of recording – especially for recordings shorter than 10 minutes in duration in combination with a minimum squared error-based matching procedure. We have shown improvements by using a maximum correlation coefficient-based matching procedure.

References

1. Union for the Co-ordination of Transmission of Electricity, Homepage, http://www.ucte.org
2. Grigoras, C.: Digital audio recording analysis – the electric network frequency criterion. International Journal of Speech Language and the Law 12(1), 63–76 (2005)
3. Kajstura, M., Trawinska, A., Hebenstreit, J.: Application of the Electrical Network Frequency (ENF) Criterion – A case of a Digital Recording. Forensic Science International 155, 165–171 (2005)
4. Cooper, A.J.: The electric network frequency (ENF) as an aid to authenticating forensic digital audio recordings – an automated approach, Conference paper. In: AES 33rd International Conference, USA (2008)
5. Brixen, E.B.: Techniques for the authentication of digital audio recordings. In: Presented at the AES 122nd Convention, Vienna (2007)

A Machine Learning Approach to Off-Line Signature Verification Using Bayesian Inference

Danjun Pu, Gregory R. Ball, and Sargur N. Srihari

Center of Excellence for Document Analysis and Recognition
Department of Computer Science and Engineering, University at Buffalo
State University of New York
Buffalo, NY, 14260, USA
srihari@cedar.buffalo.edu

Abstract. A machine learning approach to off-line signature verification is presented. The prior distributions are determined from genuine and forged signatures of several individuals. The task of signature verification is a problem of determining genuine-class membership of a questioned (test) signature. We take a 3-step, writer independent approach: 1) Determine the prior parameter distributions for means of both "genuine vs. genuine" and "forgery vs. known" classes using a distance metric. 2) Enroll n genuine and m forgery signatures for a particular writer and calculate both the posterior class probabilities for both classes. 3) When evaluating a questioned signature, determine the probabilities for each class and choose the class with bigger probability. By using this approach, performance over other approaches to the same problem is dramatically improved, especially when the number of available signatures for enrollment is small. On the NISDCC dataset, when enrolling 4 genuine signatures, the new method yielded a 12.1% average error rate, a significant improvement over a previously described Bayesian method.

1 Introduction

The automatic verification of signatures from scanned paper documents has many applications, including such tasks as the authentication of bank checks, questioned document examination, and biometrics. On-line, or dynamic, signature verification systems have been reported with high success rates [1]. However, off-line, or static research is relatively unexplored which difference can be attributed to the lack of temporal information, the range of intra-personal variation in the scanned image, etc.

We present a new approach to off-line signature verification by utilizing Bayesian methodology, in which probabilities provide a quantification of uncertainty. The writer verification problem takes as input a set of known signatures along with an unknown signature. The known signature set is partitioned into a subset of zero or more known genuine signatures and zero or more known forged signatures. The output of the writer verification problem is a boolean value as to whether or not the questioned signature is a member of the genuine signature set.

Z.J.M.H. Geradts, K.Y. Franke, and C.J. Veenman (Eds.): IWCF 2009, LNCS 5718, pp. 125–136, 2009.

Previous research has often focused on identifying random forgeries, such as that by Sabourin [2], but has less often focused on skilled forgeries. Parametric approaches such as Naive Bayes classifiers have been attempted[3], but they have an absolute requirement for having signature forgery samples as well, for training (i.e., they required the forged signature subset to have size greater than zero). Such methods viewed the problem as a two class problem (genuine and forgery) instead of a single class problem. Furthermore, this method did not learn from samples specific to an individual; in addition, with few genuine signature samples available, the estimated parameters for parametric models are likely to be incorrect and noisy. Recently [4], we described a method based on a distance based non-parametric Bayesian approach that attempted to capture the variation and similarities (distance) among the genuine signature samples and use it to classify a new questioned sample as genuine or forgery. However, this approach did not make use of prior probabilities (it assumed the prior distribution to be uniform), and therefore failed to fully capitalize on the power available to Bayesian methodology.

2   Feature Extraction

Based on a combination of large datasets consisting of many writers, we first determined the general population mean distribution for the genuine vs. genuine class and forgery vs. known class. The genuine vs. genuine class is defined as the set of statistical difference values between every genuine signatures from the an individual writer. The forgery vs. known class is the set of statistical difference values between every pair of genuine and forgery signatures. Utilizing the Bayesian parameter estimate approach, we adjust the parameters for these distributions to a particular writer. We evaluate questioned signatures by determining the probabilities according to both distributions.

2.1   Image Preprocessing

Each signature was scanned at 300 dpi gray-scale and binarized using a gray-scale histogram. Salt-and-pepper noise was removed by median filtering. Slant normalization was performed by extracting the axis of least inertia and rotating the curve until this axis coincides with the horizontal axis [5]. Given an $M \times N$ image, $G = (u_k, v_k) = (x_{(i,j)}, y_{(i,j)} | x_{(i,j)} \neq 0)$. Let S be the size of G, and let $\bar{u} = \frac{1}{S} \sum_k u_k$ and $\bar{v} = \frac{1}{S} \sum_k v_k$ be the coordinates of the center of mass of the signature. The orientation of the axis of least inertia is given by the orientation of the least eigenvector of the 2x2 matrix where $\overline{u^2} = \frac{1}{S} \sum_k (u_k - \bar{u})^2$, $\overline{v^2} = \frac{1}{S} \sum_k (v_k - \bar{v})^2$ and $\overline{uv} = \frac{1}{S} \sum_k (u_k - \bar{u})(v_k - \bar{v})$ are the second order moments of the signature [6]. The result of binarization and slant normalization of a gray-scale image is shown in Figure 1.

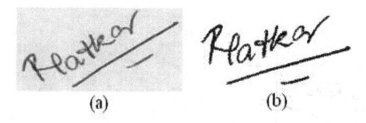

Fig. 1. Pre-processing: (a) original, (b) final

2.2 Feature Extraction

Features for static signature verification can be one of three types [7,8]: (i) global: extracted from every pixel that lie within a rectangle circumscribing the signature, including image gradient analysis [1], series expansions [9], etc., (ii) statistical: derived from the distribution of pixels of a signature, e.g., statistics of high gray-level pixels to identify pseudo-dynamic characteristics [10], (iii) geometrical and topological: e.g., local correspondence of stroke segments to trace signatures [11], feature tracks and stroke positions [7], etc. A combination of all three types of features were used in a writer independent (WI) signature verification system [12] which were previously used in character recognition [13], word recognition [14] and writer identification [15]. These features, known as gradient, structural and concavity (or GSC) features were used.

The average size of all reference signature images was chosen as the reference size to which all signatures were resized. The image is divided into a 4x8 grid from which a 1024-bit GSC feature vector is determined (Figure 2). Gradient (G) features measure the magnitude of change in a 3×3 neighborhood of each pixel. For each grid cell, by counting statistics of 12 gradients, there are $4 \times 8 \times 12 = 384$ gradient features. Structural (S) features capture certain patterns, i.e., ministrokes, embedded in the gradient map. A set of 12 rules is applied to each pixel to capture lines, diagonal rising and corners yielding 384 structural features. Concavity (C) features, which capture global and topological features, e.g., bays and holes, are $4 \times 8 \times 8 = 256$ in number.

2.3 Distance Measure

A method of measuring the similarity or distance between two signatures in feature space is essential for classification. The correlation distance performed best for GSC binary features [16] which is defined for two binary vectors X and Y as in Equation 1.

$$d(X,Y) = \frac{1}{2} - \frac{s_{11}s_{00}}{2\sqrt{(s_{10} + s_{11})(s_{01} + s_{00})(s_{11} + s_{01})(s_{00} + s_{10})}} \quad (1)$$

where s_{ij} represent the number of corresponding bits of X and Y that have values i and j. Both the WI - DS and WD - DT methods described below use $d(X,Y)$ as the distance measure.

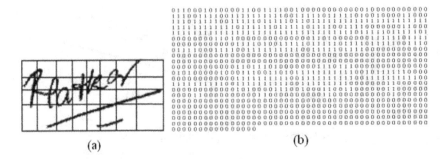

Fig. 2. Features: (a) variable grid, (b) feature vector

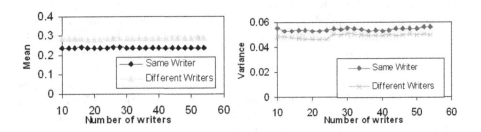

Fig. 3. Statistics of genuine-genuine and genuine-forgery distances

2.4   Distance Statistic

The verification approach of Srihari et al[15] is based on two distributions of distances $d(X,Y)$: genuine-genuine and genuine-forgery pairs. The distributions are denoted as P_g and P_f respectively. The means and variances of $d(X,Y)$ where X and Y are both genuine and X is genuine and Y is a forgery are shown in Figure 3 where the the number of writers varies from 10 to 55. Here 16 genuine signatures and 16 forgeries were randomly chosen from each subject. For each n, there are two values corresponding to the mean and variance of $n \times \binom{16}{2}$ 2 pairs of same writer (or genuine-genuine pair) distances and $n \times 16^2$ pairs of different writer (or genuine-forgery pair) distances. The values are close to constant with $\mu_g = 0.24$ and $\mu_f = 0.28$ with corresponding variances $\sigma_g = 0.055$ and $\sigma_f = 0.05$.

3   Bayesian Inference

With the statistics of genuine-genuine and genuine-forgery distance (Figure 3) in mind, one intuitive approach is to learn the mean distributions and variance distributions for different writers, considering both the genuine-genuine and genuine-forgery distances. Writer specific learning, then, focuses specifically on the writer whose identity needs to be learned from whatever known samples are available, then answering the question of whether a questioned sample was produced by this writer or not.

3.1   Mean of Genuine Writer and Genuine-Forger Distributions

If a given person has n samples, $\binom{n}{r}$ defined as $\frac{n!}{r!(n-r)!}$ pairs of samples can be compared. In each comparison, the distance between the features is computed. The result of all $\binom{n}{2}$ comparisons is a $\binom{n}{2 \times 1}$ distance vector. "genuine vs. genuine" distance values are defined as all the values in this distance vector. If a given person also has m forged samples, $m \times n$ pairs of samples can be compared (by comparing each of the genuine signatures with each of the forgeries). "forgery vs. known" distance values are defined as the results of the $m \times n$ comparisons.

For each person, then, we can generate a $\binom{n}{2}$ dimension "genuine vs. genuine" distance values vector and an $m \times n$ dimension "forgery vs. known" distance values vector. For each person, we calculate the mean values and variance values for both vectors, so that we can then generate the distribution of mean values and variance values over many different writers.

We are thus able to find the "mean of genuine vs. genuine" distribution, defined as the distribution of mean values of "genuine vs. genuine" distance values over different writers. Likewise, we can find the "mean of forgery vs. known" distribution, defined as the distribution of mean values of forgery vs. known distance values at different writers.

3.2   The Parameter Distribution

We assume that for a given person the "genuine vs. genuine" distance values are from a Gaussian distribution $N(\mu_w, \gamma_w)$ and choose the "mean of genuine vs. genuine" distribution as the conjugate prior Gaussian distribution for the parameter μ_w as $N(\mu_w|\mu 0_w, \gamma 0_w)$. The parameter $a0_w$ and $b0_w$ are hyper-parameters since they control the distribution of the parameter μ. Similarly, we chose the "mean of forgery vs. known" distribution $N(\mu_b|\mu 0_b, \gamma 0_b)$ as the conjugate prior Gaussian distribution for the "forgery vs. known" distance value $N(\mu_b, \gamma_b)$. In this paper, we just consider the distribution of parameter μ and assume parameter γ is uniform (future work involves relaxing both parameters μ and γ).

3.3   Algorithm

The algorithm proceeds through three steps: (i) build the prior "mean of genuine vs. genuine" distribution and the "mean of forgery vs. known" distribution from the general population datasets, (ii) train genuine and forgery sets by adjusting prior parameter distribution, and (iii) calculate probabilities of model membership.

Step 1: Extract Prior Knowledge. In our general population datasets, we have several writers and each writer has several genuine samples and forged samples. Using the method introduced above, we obtain the "mean of genuine vs. genuine" distribution $N(\mu_b|\mu 0_w, \gamma 0_w)$ and the 'mean of forgery vs. known" distribution $N(\mu_b|u 0_b, \gamma 0_b)$, where $u0_w$ and $u0_b$ are the mean parameters; $\gamma 0_w$ and $\gamma 0_b$ are the precision parameters.

Here we combine all the "genuine vs. genuine" distance value vectors from all the writers to a large, general population genuine-genuine distance dataset and also combine all "forgery vs. known" distance value vectors from all the writers to generate a general population forgery-genuine distance dataset. For the genuine-genuine distance dataset, we assumed it to be a Gaussian distribution $N(u_w, \gamma 1_w)$ and similarly, for the forgery-genuine distance dataset, $N(u_b, \gamma 1_b)$, where u_w and u_b are the mean parameters; $\gamma 1_w$ and $\gamma 1_b$ are the precision parameters. Furthermore, we assume that $\gamma 0_w = b0_w * \gamma 1_w$ and $\gamma 0_b = b0_b * \gamma 1_b$.

We define the hyperparameter for the parameter distribution as $a0_w = \mu 0_w$, $b0_w = \frac{\gamma 0_w}{\gamma 1_w}$, $a0_b = \mu 0_b$, and $b0_b = \frac{\gamma 0_b}{\gamma 1_b}$.

Step 2: Create Posterior Distribution. In this training phase, for a specific person, we enroll m genuine signatures ($m \geq 0$) and n ($n \geq 0$) forged signatures. We obtain the "genuine vs. genuine" distance values D_w and "forgery vs. known" distance values D_b and their likelihood function $N(\mu_w, \gamma_w)$ and $N(\mu_b, \gamma_b)$ for this particular writer.

By the Bayesian Rule, we obtain:

$$N(\mu_w|D_w, a_w, b_w * \gamma_w) \propto N(D_w|\mu_w, \gamma_w) \times N(\mu_w|a0_w, b0_w * \gamma_w) \qquad (2)$$

$$N(\mu_b|D_b, a_b, b_b * \gamma_b) \propto N(D_b|\mu_b, \gamma_b) \times N(\mu_b|a0_b, b0_b * \gamma_b) \qquad (3)$$

We normalize them, and can define their hyper parameters as $a_w = \frac{m * \overline{D_w} + a0_w * b0_w}{m + b0_w}$, $b_w = \frac{1}{m + b0_w}$, $a_b = \frac{n * \overline{D_b} + a0_b * b0_b}{n + b0_b}$, and $b_b = \frac{1}{n + b0_b}$, where m is the cardinality of D_w, n is the cardinality of D_b, $\overline{D_w}$ is the mean value of the distance values in D_w, and $\overline{D_b}$ is the mean value of distance values in D_b.

Step 3: Sum and Product Rules of Probabilities. When evaluating a questioned signature, we compare it to the m genuine enrolled signatures and get a distance dataset $X = \{x_1, ..., x_m\}$. The distance x is assumed to be normal, $N(x|\mu, \gamma)$, where μ and γ are the mean and precision of the normal distribution. So the probability of data set X can expressed as:

$$P(X|\mu, \gamma) = \Pi_{i=1}^{m}(\frac{\gamma}{2 * \pi})^{\frac{1}{2}} e^{-\frac{\gamma * (X_i - \mu)^2}{2}} = (\frac{\gamma}{2 * \pi})^{\frac{m}{2}} e^{-\frac{m * S * \gamma}{2}} e^{-\frac{m * \gamma * (\mu - \overline{X})^2}{2}} \qquad (4)$$

where S is the variance of dataset X.

And the probability can be expressed as the marginalization:

$$P(X) = \int_{0}^{\infty} \int_{-\infty}^{\infty} P(X, \mu, \gamma) d\mu d\gamma \qquad (5)$$

$$= \int_{0}^{\infty} \int_{-\infty}^{\infty} P(X|\mu, \gamma) P(\mu, \gamma) d\mu d\gamma \qquad (6)$$

$$= \int_{0}^{\infty} \int_{-\infty}^{\infty} P(X|\mu, \gamma) P(\mu|\gamma) P(\gamma) d\mu d\gamma \qquad (7)$$

where $N(\mu|\gamma)$ and $P(\gamma)$ are the prior conjugate Normal and Gamma distribution for parameter μ and γ respectively, described as:

$$P(\mu|\gamma) = N(\mu|a, b*\gamma) = \left(\frac{b*\gamma}{2*\pi}\right)^{\frac{1}{2}} * e^{-\frac{b*\gamma*(\mu-a)^2}{2}} \tag{8}$$

$$P(\gamma) = P(\gamma|c, d) = \frac{d^c * \gamma^{c-1} * e^{-d*\gamma}}{\Gamma(c)} \tag{9}$$

If we assume the distribution of parameter γ is uniform, we can get:

$$P(X) = \int_0^\infty \int_{-\infty}^\infty P(X|\mu, \gamma) \times N(\mu|a, b*\gamma)d\mu d\gamma. \tag{10}$$

substitute equations 4 and 8, we can get:

$$P(X) = \int_0^\infty \int_{-\infty}^\infty \left(\frac{\gamma}{2*\pi}\right)^{\frac{m}{2}} e^{-\frac{m*S*\gamma}{2}} e^{-\frac{m*\gamma*(\mu-\overline{X})^2}{2}} * \left(\frac{b*\gamma}{2*\pi}\right)^{\frac{1}{2}} * e^{-\frac{b*\gamma*(\mu-a)^2}{2}} d\mu d\gamma \tag{11}$$

$$= \left(\frac{b}{m+b}\right)^{\frac{1}{2}} \frac{1}{(2\pi)^{\frac{m}{2}}} \frac{\Gamma(Z)}{A^Z} \tag{12}$$

where $A = \frac{mS}{2} + \frac{mb(a^2+\overline{X^2})}{2(m+b)}$; $Z = \frac{m}{2} + 1$ and S is the variance of data set X.

We first suppose the questioned signature belongs to genuine class:

$$P_w(X) = \int_0^\infty \int_{-\infty}^\infty P(X|\mu, \gamma) \times N(\mu|a_w, b_w * \gamma)d\mu d\gamma \tag{13}$$

According to Equation 12, we get

$$P_w(X) = \left(\frac{b_w}{m+b_w}\right)^{\frac{1}{2}} \frac{1}{(2\pi)^{\frac{m}{2}}} \frac{\Gamma(Z)}{A^Z}; \tag{14}$$

where $A = \frac{mS}{2} + \frac{mb_w(a_w^2+\overline{X^2})}{2(m+b_w)}$.

We then suppose the questioned signature belongs to forgery class:

$$P_b(X) = \int_0^\infty \int_{-\infty}^\infty P(X|\mu, \gamma) \times N(\mu|a_b, b_b * \gamma)d\mu d\gamma \tag{15}$$

According to Equation 12, we get

$$P_b(X) = \left(\frac{b_b}{m+b_b}\right)^{\frac{1}{2}} \frac{1}{(2\pi)^{\frac{m}{2}}} \frac{\Gamma(Z)}{A^Z}; \tag{16}$$

where $A = \frac{mS}{2} + \frac{mb_b(a_b^2+\overline{X^2})}{2(m+b_b)}$

Decision. We compare the posterior class probabilities $P_w(X)$ and $P_b(X)$ to achieve a decision. If $P_w(X) \geq P_b(X)$, the algorithm asserts that the questioned signature is a genuine signature, otherwise it asserts the questioned signature is a forgery.

4   Experiment and Performance

4.1   Datasets

We used three datasets available to us for training and experiments.

The first dataset (the CEDAR dataset) was originally presented in [12], which contains 55 individuals, each with 24 genuine signatures and 24 forgeries forged by 3 other writers. One example of each of 55 genuine are shown in Figure 4.

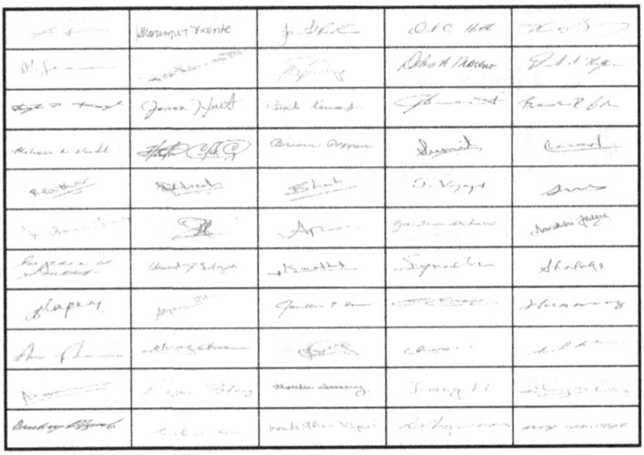

Fig. 4. Genuine signature samples from CEDAR dataset

Each signature was scanned at 300 dpi gray-scale and binarized using a gray-scale histogram. Salt pepper noise removal and slant normalization were two steps involved in image preprocessing.

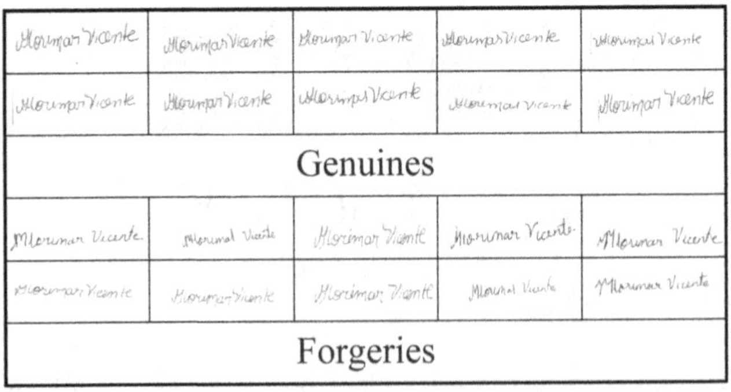

Fig. 5. Samples for writer No.21

Table 1. Number of signatures for each category

	John Moll	Marie Wood
Normal	140	179
Disguised	35	21
Spurious	90	0
Simulated	449	650
Total	714	850

The second dataset used (the Georgia dataset) was originally created by the American Board of Forensic Document Examiners (ABFDE), and the purpose of it is for academic research and FDE professional development and training. It contains 1564 genuine and forgery signatures of two specimens: "John Moll" and "Marie Wood". Both genuine and forgery have two sub-categories, for genuine, there are normal and disguised signatures; and for forgery, there are spurious and simulated signatures. Spurious signatures are naturally written signatures written by one writer in the name of another writer; no attempt to disguise, distort or at simulating the writing of another. Table 1 shows the number of signatures for each category. For simulated signatures, "John Moll" and "Marie Wood" were simulated by 9 and 13 individuals respectively. In addition, genuine signatures were written in 7 continuous days while simulated samples were created in 5 continuous days. All the signature samples were scanned at 300 dpi gray-scale. One sample image for each category is shown in Figure 6.

The third dataset (the NISDCC dataset) used was the participant dataset used for the ICDAR 2009 signature verification competition[17], consisting of 1920 images binarized at 300dpi. Twelve writers created five authentic signatures. Each writer had a corresponding forgery set consisting of five imitation signatures written by 31 individuals (155 forged signatures total for each of the 12).

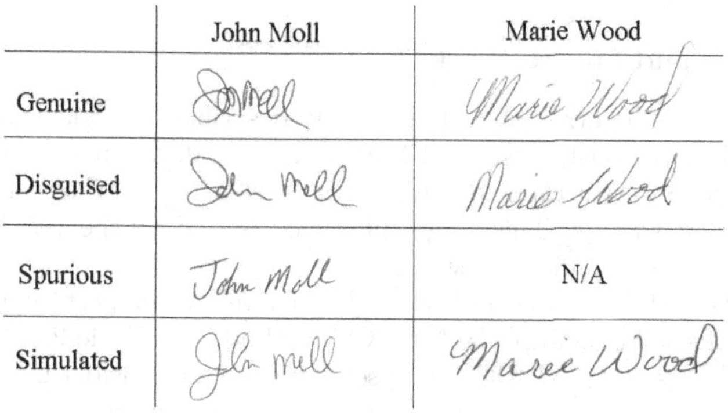

Fig. 6. Sample signature images for each category

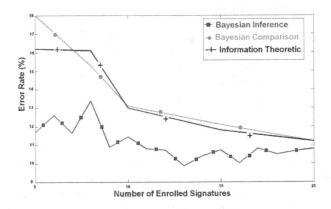

Fig. 7. Error Rate on Georgia dataset vs. Number of Enrolled Signatures

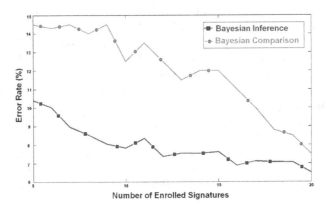

Fig. 8. Error Rate on CEDAR dataset vs. Number of Enrolled Signatures

4.2   Results and Future Work

When comparing the error rate of the newly described method against the error rate of the previously described Bayesian method, the new method always achieves a superior error rate. The difference is especially marked for lower numbers of enrolled writers. This large difference is due to the newly described method making use of prior knowledge.

The results of the newly described method as well as the performance of some of our other methods against the NISDCC method is in Table 2.

Although we used the NISDCC data set, since we are considering a fundamentally different problem, the different results are not comparable. In the competition system, we just calculated the feature distance between two signatures; however, in the problem we consider, we actually model each individual signature, as well as utilize prior knowledge. Any approach which generates comparable

Table 2. NISDCC results

Method	False Reject %	False Accept %	Avg. Error Rate
Graph Matching	25.0	50.2	37.6
Information-Theoretic	20.0	30.3	25.2
Bayes-Comparison	21.7	25.4	23.6
Bayes Inference	8.4	15.8	12.1

distances between signatures should achieve the same significant improvement that we experienced here. Since this method is a general methodology which can be applied with other signature feature sets, we will try to apply this methods to others in the future.

To accomplish this rate for the other three older methods, a threshold needed to be determined. However, for the new method, a threshold is not used because, for a particular writer, the mean value of the "forgery vs. known" distance value is much larger than the mean value of the "genuine vs. genuine" distance value. In addition, the results were nearly identical whether we built the prior knowledge from the CEDAR signature set alone, the CEDAR set along with the Georgia dataset, or all three sets. The reason for this was that the "genuine vs. genuine" distance distribution and "forgery vs. known" distance distribution changes only slightly when the prior knowledge data set increases. This method also scales with all available data. While forgery signatures are not required for it to work, they can be incorporated into the model if available to enhance the quality of the results.In NISDCC dataset, when no forgery signature is enrolled, the error rate is about 12% which is higher than the error rate when several forgery signatures are enrolled.

In the future, we are interested in considering additional variations to the Bayesian formulation, such as considering the variance along with the mean value of the two prior distributions.

References

1. Sabourin, R., Plamondon, R.: Preprocessing of handwritten signatures from image gradient analysis. In: Proceedings of the 8th International Conference on Pattern Recognition, pp. 576–579 (1986)
2. Sabourin, R., Genest, G., Prteux, F.J.: Off-line signature verification by local granulometric size distributions. IEEE Transactions on Pattern Analysis and Machine Intelligence 19, 976–988 (1997)
3. Srihari, S.N., Xu, A., Kalera, M.K.: Learning strategies and classification methods for off-line signature verification. In: Proceedings of the 7th International Workshop on Frontiers in handwriting recognition (IWHR), pp. 161–166 (2004)
4. Srihari, S.N., Kuzhinjedathu, K., Srinivasan, H., Huang, C., Pu, D.: Signature verification using a bayesian approach. In: Srihari, S.N., Franke, K. (eds.) IWCF 2008. LNCS, vol. 5158, pp. 192–203. Springer, Heidelberg (2008)
5. Horn, B.: Robot vision. MIT Press (1986)

6. Munich, M.E., Perona, P.: Visual identification by signature tracking. IEEE Transactions on Pattern Analysis and Machine Intelligence 25(2), 200–217 (2003)
7. Fang, C.H.L., Tang, Y.Y., Tse, K.W., Kwok, P.C.K., Wong, Y.K.: Off-line signature verification by the tracking of feature and stroke positions. Pattern Recognition 36, 91–101 (2003)
8. Lee, S., Pan, J.C.: Off-line tracing and representation of signatures. IEEE Transactions on Systems, Man and Cybernetics 22, 755–771 (1992)
9. Lin, C.C., Chellappa, R.: Classification of partial 2-d shapes using fourier descriptors. IEEE Trans. Pattern Anal. Mach. Intell. 9(5), 686–690 (1987)
10. Ammar, M., Yoshido, Y., Fukumura, T.: A new effective approach for off-line verification of signatures by using pressure features. In: Proceedings of the 8th International Conference on Pattern Recognition, pp. 566–569 (1986)
11. Guo, J.K., Doermann, D., Rosenfeld, A.: Local correspondence for detecting random forgeries. In: Proceedings of the International Conference on Document Analysis and Recognition, pp. 319–323 (1997)
12. Kalera, M.K., Srihari, S., Xu, A.: Off-line signature verification and identification using distance statistics. International Journal of Pattern Recognition and Artificial Intelligence, 228–232 (2003)
13. Srikantan, G., Lam, S., Srihari, S.: Gradient-based contour encoding for character recognition. Pattern Recognition 7, 1147–1160 (1996)
14. Zhang, B., Srihari, S.N.: Analysis of handwriting individuality using handwritten words. In: Proceedings of the Seventh International Conference on Document Analysis and Recognition. IEEE Computer Society Press, Los Alamitos (2003)
15. Srihari, S.N., Cha, S., Arora, H., Lee, S.: Individuality of handwriting. Journal of Forensic Sciences 47(4), 858–872 (2002)
16. Zhang, B., Srihari, S.N.: Binary vector dissimilarity measures for handwriting identification. In: Proceedings of SPIE, Document Recognition and Retrieval, pp. 155–166 (2003)
17. Blankers, V., van den Heuvel, C., Franke, K., Vuurpijl, L.: The icdar 2009 signature verification competition. In: Tenth International Conference on Document Analysis and Recognition (ICDAR 2009) (2009)

Computer-Assisted Handwriting Analysis: Interaction with Legal Issues in U.S. Courts

Kenneth A. Manning[1] and Sargur N. Srihari[2]

[1] Phillips Lytle LLP, Buffalo, NY 14203
kmanning@phillipslytle.com
[2] University at Buffalo, Amherst, NY 14228
srihari@cedar.buffalo.edu

Abstract. Advances in the development of computer-assisted handwriting analysis have led to the consideration of a computational system by courts in the United States. Computer-assisted handwriting analysis has been introduced in the context of Frye or Daubert hearings conducted to determine the admissibility of handwriting testimony by questioned document examiners, as expert witnesses, in civil and criminal proceedings. This paper provides a comparison of scientific and judicial methods, and examines concerns over reliability of handwriting analysis expressed in judicial decisions. Recently, the National Research Council assessed that "the scientific basis for handwriting comparisons needs to be strengthened". Recent studies involving computer-assisted handwriting analysis are reviewed in light of the concerns expressed by the judiciary and National Research Council. A future potential role for computer-assisted handwriting analysis in the courts is identified.

1 Introduction

The resolution or accommodation of factual uncertainty is essential to the process of decision-making, whether the decision relates to research in a scientific or engineering discipline or to a determination of a dispute in a judicial proceeding. In scientific and engineering applications, the scientific method is the fundamental tool for testing hypotheses, resolving uncertainty and developing knowledge in a rigorous manner. Where mathematical models are utilized, newer computational methods are being applied to probability analysis, yielding greater capacity for data mining, pattern recognition and predictive accuracy.

In the United States, the Judicial Branch of Government concerns itself with the resolution of both legal and factual uncertainty, through decisions made by Judges or verdicts rendered by juries. The fundamental mechanism for resolving uncertainty in the judicial system is the introduction of evidence for consideration by the court or jury, the cross-examination of witnesses through whom the evidence is introduced, and the rendering of a decision by the Court or a verdict by the jury.

The differing methods for testing hypotheses in a scientific or engineering setting and for resolving uncertainty in a judicial forum create sharp conflict when scientific or engineering evidence is attempted to be introduced in a courtroom. A primary

Z.J.M.H. Geradts, K.Y. Franke, and C.J. Veenman (Eds.): IWCF 2009, LNCS 5718, pp. 137–149, 2009.

example of such a conflict is presented by evidence relating to handwriting analysis. This paper will examine differences in approach in scientific and judicial models, the standards for admissibility of scientific or technical methods in federal courts and in a representative state court (New York). It will then examine concerns expressed by the courts over handwriting analysis and the recent assessment of the National Research Council that the scientific basis for handwriting comparisons needs to be strengthened. Finally, it will review advances in the development of computer-assisted handwriting analysis and consider the impact and potential future use of computer-assisted handwriting analysis in the courts.

2 Scientific and Judicial Models: Different Approaches

The principles of science and interpretation of scientific data are concisely described by the scientific method. The National Research Council has observed: " The way in which science is conducted is distinct from, and complementary to, other modes by which humans investigate and create. The methods of science have a long history of successfully building useful and trustworthy knowledge and filling gaps while also correcting past errors. The premium that science places on precision, objectivity, critical thinking, careful observation and practice, repeatability, uncertainty management, and peer review enables the reliable collection, measurement, and interpretation of clues in order to produce knowledge."[1].

The Judicial model in the United States (which differs from European models), creates a highly structured environment in an adversarial setting, where advocates (i.e., trial attorneys) present opposing positions in a dispute to be resolved by a judge or jury. The U. S. judicial model relies heavily on precedent, tending to preserve the status quo. It does not afford a comfortable environment for scientific experimentation or acknowledgment of weaknesses or limitations in methodology. The information heard and seen by the Court, for presentation and consideration by a trial jury, comes in the form of evidence. For evidence to be admissible in the face of an objection by the adversary, it must have sufficient authenticity, foundation, and evidentiary basis under governing procedural and substantive rules to permit the Court to rule in favor of its admissibility in evidence and to permit its consideration by the jury.

To challenge or test the strength of evidence in the courtroom, cross-examination is permitted. Cross-examination has been described by those within the U.S. judicial system as "... beyond any doubt the greatest legal engine ever invented for the discovery of truth".[2] Given leeway to ask probing and leading questions, a defense attorney is permitted to explore the factual basis for testimony, to challenge a witness for bias or favor, and to impeach the credibility of a witness on a variety of grounds.

3 Standards for Admissibility of Expert Testimony

The modern development of the law of evidence relating to testimony of scientific or technical experts was guided by a decision of the Court of Appeals for the D. C. Circuit in a criminal case in 1923, Frye v. United States [3] (Frye). The court in Frye set forth the standard for admissibility of expert testimony: "Just when a scientific principle or discovery crosses the line between the experimental and demonstrable states is difficult to define. Somewhere in this twilight zone the evidential force of the principle must be recognized, and while courts will go a long way in admitting expert testimony deduced from a well-recognized scientific principle or discovery, the thing from which the deduction is made must be sufficiently established to have gained general acceptance in the particular field in which it belongs".

Many state and federal trial and appellate courts found the decision in Frye to be persuasive, and either formally adopted or simply followed the so-called "Frye test" for determining the admissibility of scientific testimony based on general acceptance in the particular field in which it belongs. New York State courts, for example, generally have followed the Frye rule. Under Frye, the trial judge does not determine whether a novel scientific methodology is actually reliable. Rather, the judge determines whether most scientists in the field believe it to be reliable. If so, then the methodology is reliable enough for the jury to consider.

The Frye test has engendered significant controversy, even in New York, where it is followed. The criticism of the test may be summarized as follows: (1) it represents an overly conservative approach to admissibility that excludes reliable expertise simply because it has not been in existence long enough to be generally accepted by the scientific community; (2) its vagueness makes the standard subject to manipulation; (3) the identification of the appropriate discipline or professional field can be problematic, because many scientific techniques cut across several disciplines; (4) the percentage of experts in a field that must accept a technique before it can be accorded the status of "general acceptance" is unclear; and (5) the delay between the introduction of a scientific technique and its recognition as generally accepted prevents reliable evidence from being heard in court [4].

In 1993, the United States Supreme Court decided in Daubert v. Merrell Dow Pharm, Inc.[5] (Daubert), a civil case, that the provisions of Federal Rule of Evidence 702, rather than the common law rule developed in Frye, governed admissibility of expert testimony in federal courts. In reviewing the admissibility of scientific testimony, the Court observed that "evidentiary reliability will be based upon scientific validity" and directed trial judges to ensure the reliability of all scientific testimony and evidence, by focusing on the expert's principles and methodology. The Court listed five (non-exclusive) factors to be given consideration by the trial court in making a ruling on the admissibility of expert testimony: (1) whether a theory or technique can be (and has been) tested; (2) whether the theory or technique has been subjected to peer review and publications; (3) the known or potential rate of error of a particular scientific technique; (4) the existence and maintenance of standards controlling the technique's operation; and (5) a scientific technique's degree of acceptance within a relevant scientific community. The subsequent decision of the United States Supreme Court in Kumho Tire Co., v. Carmichael [6] brought technical, as well as scientific, expert testimony within the ambit of Rule 702.

4 Evidence of Handwriting Analysis – Concerns over Reliability

Handwriting is the product of an individual's neuromuscular function applied to an instrument or implement to create ordered markings, usually in a particular language, on a receptive medium or surface, commonly paper. Accordingly, the handwriting reflects the genetic, biological, and structural aspects of the writer, as well as the writer's environment, experience and training. Questioned document examination involves the comparison and analysis of documents and printing and writing instruments in order to identify or eliminate persons as the source of the handwriting. Questions about documents arise in research, business and finance, as well as civil and criminal trials. Questioned document examination includes the field of handwriting identification.

The examination of handwritten items typically involves the comparison of a questioned item submitted for examination along with a known item of established origin associated with the matter under investigation. Comparisons are based on the high likelihood that no two persons write the same way, while considering the fact that every person's writing has its own variabilities. "Thus, an analysis of handwriting must compare interpersonal variability -- some characterization of how handwriting features vary across a population of possible writers -- with intrapersonal variability -- how much an individual's handwriting can vary from sample to sample. Determining that two samples were written by the same person depends on showing that their degree of variability, by some measure, is more consistent with intrapersonal variability than with interpersonal variability..."[7]

Expert opinions offered by questioned document examiners combine elements of both objectivity and subjectivity. Terminology has been developed for expressing the conclusions and opinions of handwriting comparison and identification. Several numerical scales have been developed among document examiners in various countries to express to courts, juries, and others, the degree of confidence in an opinion. A nine-point scale is frequently used by questioned document examiners worldwide, and provides a metric for strength of opinion, ranging from definite conclusions to weaker "indications".

In New York, the admissibility of opinion testimony of handwriting experts has been settled [8]. In 1978, New York's highest court adopted the following statement of the law, in the context of a criminal case: "It ought to be now well understood that the identification and the decipherment of documents, including handwriting and all other features, are the subject of scientific study, made by the aid of instruments of precision, and that modern research has elevated the whole subject into the realm of an applied science. Under such conditions, tangible reasons can be given for every opinion, precisely as an engineer can explain the reasons for constructing a bridge of a certain type. A qualified expert's opinion may therefore now be tested and judged by the reasons on which it is based (citation omitted) "[9]. Testimony of handwriting analysis has been regularly admitted in New York courts as satisfying the Frye test of general acceptance.

New York State sits within the geographic federal jurisdiction of the Second Circuit Court of Appeals. The Second Circuit has observed that the shift expressed in Daubert to a more permissive approach to expert testimony in federal courts did not abrogate the trial court's function as a gatekeeper of reliable evidence [10]. More importantly, the Second Circuit has observed that the departure, under Federal Rule 702, from the Frye standard, did not "grandfather", or protect from Daubert scrutiny, evidence that had been previously admitted under Frye [11].

There now have been reported decisions in many federal district courts, and seven of the thirteen federal Circuit Courts of Appeal, involving the admissibility of handwriting testimony since <u>Daubert</u> was decided; most have been criminal cases. Some have been decided on paper submissions, such as affidavits; others involved pretrial <u>Daubert</u> hearings; and others, rulings on evidence at trial. The seven Circuit Courts of Appeal that have ruled are consistent in their position that the methods employed by qualified questioned document examiners generally are reliable, and that handwriting testimony generally is admissible. The results, in a given case at the District Court (trial court) level, are more variable, and reflect rulings on proof concerning not only the admissibility of handwriting expert methodology, but also the qualifications of the expert, and whether the methodology was reliably followed. Analysis of published cases shows that some trial courts have chosen to exclude the more powerful testimony of authorship of a questioned document, while admitting testimony showing similarities and differences between questioned and known documents.

Most of the federal courts have decided that the methodology used for handwriting analysis is sufficiently reliable to admit testimony of qualified questioned document examiners based on adequate samples of known and questioned documents. Nevertheless, an examination of the judicial decisions involving more fully developed records, where handwriting testimony was excluded, in whole or in part, and the judicial decisions where dissenting viewpoints were expressed, is helpful to understand the differences in outcomes, as well as the judicial concerns over the reliability of handwriting analysis.

The judicial concerns over the reliability of handwriting analysis can be summarized as follows: (1) fundamental premises – there has been little empirical testing and few published studies relating to the basic theories on which handwriting analysis is based; (2) reliability of methodology used by questioned document examiners – the data on handwriting analysis is "sparse, inconclusive, and highly disputed, particularly as regards the relative abilities of [questioned] document examiners and lay examiners", [12] and "[prior studies] did not conclusively establish that [questioned document examiners] can reliably do what they say they can do" [13]; (3) error rates – there is little known about the error rates of questioned document examiners, and the error rates for "real world conditions" are higher [14]; (4) peer review and publications – peer review has been limited to the questioned document examiner community and those within the field of questioned document examination have failed to engage in critical study of the basic principles and methods of handwriting analysis; published articles on handwriting analysis are "significantly different from scholarly articles in such fields as medicine or physics in their lack of critical scholarship" [15]. There has been no peer review by "a competitive, unbiased community of practitioners and academics" [16]; and (5) general accepance – the acceptance of the methodology should come from disinterested experts outside of the questioned document examiner community.

The National Research Council of the National Academies prepared a report entitled <u>Strengthening Forensic Science in the United States: A Path Forward.</u> The report charts a path for the strengthening of forensic science, when viewed from the perspective of the United States judicial system. Numerous hearings were held, at which presentations were made concerning a number of forensic sciences, including questioned document examination. The committee and presenters included a cross-section of questioned document examiners, scientists, professors, and other professionals, a number of whom have either testifed at <u>Daubert</u> hearings or at trial in reported federal cases, or whose work has been cited in those cases. The Summary Assessment of the Committee relative to handwriting analysis is concisely stated in the report: "The scientific basis for handwriting comparisons needs to be strengthened. Recent studies have increased our understanding of the individuality and consistency of handwriting and computer studies and suggest that there may be a scientific basis for handwriting comparison, at least in the absence of intentional obfuscation or forgery. Although there has been only limited research to quantify the reliability and replicability of the practices used by trained document examiners, the Committee agrees that there may be some value in handwriting analysis" [17].

5 The Development of Computer-Assisted Handwriting Analysis

The concerns over the reliability of handwriting analysis that have been expressed by at least some members of the judiciary reflect the circumstance that the admissibility or weight given to an expert opinion may be strongly based on the credibility and professional standing of the questioned document examiner, rather than on the scientific basis supporting his or her opinion. To the extent that scientifically acceptable methods can be used to develop better data or reliably assist in the process of handwriting analysis, at least some of those concerns can be addressed.

Over the past thirty years, research has been conducted relative to using computers to enhance or automate the handwriting analysis conducted by questioned document examiners. [18] Because computer assisted systems are based on measurements taken from two-dimensional documents or electronically scanned images, they do not permit evaluation of all of the attributes of handwriting considered by questioned document examiners. Nevertheless, the availability of computational systems brings the potential for more objective spatial feature measurements, either to assist in the formulation of an opinion by a questioned document examiner, or to directly yield comparative information relating to questioned and known samples. Research on objective measurement strategies to assist questioned document examiners in making judgments about spatial and feature comparisons has now accelerated.

A matrix analysis technique described by Found [19] compared measurements to generate a spatial consistency score in order to determine whether a questioned image was consistent or inconsistent with the range of variation in a standard image group. Questioned document examiners could then use the information to explore hypotheses from which opinions regarding authorship could be given. Found concluded that "Ultimately we are moving toward systems which would employ technology such as neural networks to predict whether questioned signatures are genuine or simulated. Such a technique could then be subjected to validation trials and the error rate calculated. This type of objectivity in forensic science forms the future goal for research of this type and offers considerable promise to the field of forensic handwriting examination..."

More recently, computer scientists have begun applying computer-vision and pattern recognition techniques to problems relating to writer identification [20]. A

2003 survey of computer methods in questioned document examination reviewed key techniques and published results. [21] The survey reported on more recent efforts at developing a computational theory of handwriting analysis; i.e., "an application of computer vision which consists of representations, algorithms and implementations". From a scientific viewpoint, a computational theory approach has the advantage of repeatability. The same results may be obtained for the same documents, as compared to the subjectivity and potentially differing opinions of expert human document examiners.

A computational theory of handwriting examination was developed and the theory tested with experiments for handwriting identification, as well as for handwriting verification (the 2002 study) [22]. The experiments were conducted in response to judicially expressed concerns over the reliability of handwriting analysis. The premise for the experiments was that a writer's individuality of handwriting rests upon the hypothesis that each individual has consistent handwriting that is distinct from the handwriting of another individual. The study expressly acknowledged that this hypothesis has not been subjected to rigorous scrutiny with the accompanying experimentation, testing, and peer review. The technology for the study was built on recent advances in machine-learning algorithms for recognizing handwriting on electronically-scanned documents. The goal of the study was to establish, in an objective manner, the intuitive observation that the within writer variance (the variation within a person's handwriting samples) is less than the between-writer variance (the variation between the handwriting samples of two different people).

The data collection was comprised of obtaining three full pages of handwriting, based on a designed source document, from a randomly selected population (approximately 1500 individuals), large enough to enable properly drawn inferences about the general population. The writer population for the study was designed so that it would be representative of the United States population. From the samples, both macro features obtained at the global level from the document, and micro features, obtained at the level of individual handwriting characters, were utilized. Feature extraction was performed to obtain handwriting attributes that would enable the writing style of one writer to be discriminated from the writing style of another writer. The features were electronically scanned and converted into digitized images. A binarization algorithm was used to convert the images into binary images. Paragraph, line, and word images were segmented for analysis. Once the images were segmented, macro and micro computational features (those that can be determined algorithmically by software operating on the scanned image of the handwriting) were extracted.

Computational features remove subjectivity from the process of feature extraction. The computational features for which algorithms were used in the study included 11 macro features, which are loosely related to conventional features [23] used by questioned document examiners. Micro features were computed at the allograph, or character shape, level. They are analogous to the allograph-discriminating elements among questioned document examiner features. The micro features used for the experiment included gradient, structural and concavity (GSC) attributes, which are those used in automatic character recognition processes.

Once the macro and micro features were extracted, both identification and verification models were formed as tasks for machine-learning algorithms. Both models involved a method of measuring similarity. The writer identification model used features extracted from images and the learning algorithm known as "The Nearest Neighbor Rule". Using this algorithm, on both macro and micro features, the writer identification model resulted in a relatively high accuracy rate (the correct writer was identified in approximately 98% of the cases when all possible pairs of writers were considered). The verification model used a mapping from feature space into distance space. The feature distances were compared using a machine-learning techique based on artificial neural networks to classify the between- and within-writer distances. Using this algorithm, an accuracy of 95% was achieved, at the document testing level. Lower degrees of accuracy were reported at the paragraph, word, and character level. The study concluded that the objective analysis that was done should provide the basis for the conclusion of individuality when a human analyst is measuring the finer features by hand. The study also noted that there were important extensions of the work that could be done, including the study of handwriting of similarly trained individuals, and the study of variations of handwriting over periods of time.

6 Judicial Consideration of the Computer-Assisted Handwriting Analysis in Daubert Hearings

Pretrial evidentiary hearings in federal court (Daubert hearings) are governed by both the Federal Rules of Evidence, as well as judicial decisions. Under Federal Rule of Evidence 104, "Preliminary questions concerning the qualification of a person to be a witness, … or the admissibility of evidence shall be determined by the court … . In making its determination it is not bound by the Rules of Evidence…" [24]. Accordingly, in Daubert hearings, the formal rules of evidence do not control the information considered by the Court in determining, in advance, whether to admit testimony or other evidence at trial [25].

The computer-assisted writer identification and verification experiments reported in the 2002 study have been reviewed by courts in the context of Daubert hearings on several occasions. In U.S. v. Prime, the Ninth Circuit Court of Appeals discussed the experiments in ruling on a motion by a criminal defendant to exclude trial testimony of a questioned document examiner as unreliable. In determining whether handwriting analysis theory or technique can be tested, the Court observed, "… The Government and [questioned document examiner] provided the court with ample support for the proposition that an individual's handwriting is so rarely identical that expert handwriting analysis can gauge reliably the likelihood that the same individual wrote two samples. The most significant support came from Professor Sargur N. Srihari of the Center of Excellence for Document Analysis and Recognition at the State University of New York at Buffalo, who testified that the result of his published research was that 'handwriting is individualistic'. … [26].

More recently, in a 2007 District Court decision in U.S. v.Yagman [27], a questioned document examiner was permitted to testify as to authorship of disputed documents, and the Court considered the 2002 study in connection with the known or potential error rate of handwriting analysis. Adopting the discussion of the 2002 study reflected in the District Court decision in U.S. v. Gricco [28], the court observed

" ... the state of the art of handwriting analysis has improved and progressed. The more recent studies show the reliability of handwriting analysis. Most recently, document analysts conducted a study in which they undertook to validate the hypothesis that handwriting is individualistic and that handwriting is reliable. [Citing to the 2002 study]". Accordingly, federal courts have been receptive to considering the results of computer-assisted handwriting analysis experiments in the context of Daubert hearings, conducted for the purpose of admitting or excluding testimony of questioned document examiners at trial. The results have been received by the courts to support the testimony even though the computer-assisted analysis was not performed by the questioned document examiners themselves.

7 Further Developments in Computational Systems

Since the 2002 study, a more complete system for handwriting examination has been developed using computational theory. [29] CEDAR-FOX [30] is a computer-based system for analyzing electronically scanned handwriting documents, and searching electronically stored repositories of scanned documents. The system is primarily designed for questioned document examination, and it has a number of functionalities which make it useful for analyzing documents or searching handwritten notes and historical manuscripts. Because the system relies on electronic scanning of original documents, it does not have the capability of analyzing the chemical or physical properties of the medium of the original document, and cannot directly measure any aspect of the handwriting that may be three-dimensional in nature, such as indentations. The performance of the system depends, in part, on the quality of the scanned images.

A significant feature of CEDAR-FOX is the incorporation of machine-learning and pattern recognition techniques. Based on the availability of multiple samples of known handwriting, the system analyzes the known handwriting by making spatial measurements for selected features, and then mapping those measurements in the form of binary vectors into computer distance space. By measuring, assembling and comparing binary feature vector data, the system learns through training on the known samples, thereby accommodating natural variations in handwriting among known samples. Central to both identification and verification is the need for associating the quantitative measure of similarity between two samples. Such a quantitative measure brings both an assurance of repeatability and a degree of objectivity. By using the same types of measurements and assemblage of data for one or more questioned documents, the system provides a means for comparison, based on accepted theories of probability, of the distributions of data of both the known and questioned documents. The mathematical probability resulting from the comparisons is expressed in the form of a logarithmic likelihood ratio (LLR). Positive correlations indicate sameness of authorship. Negative correlations indicate difference in authorship. The magnitude of the LLR indicates the strength of the statistical correlation, and the relative confidence in the indication.

The CEDAR-FOX system, as currently designed, does not replace traditional handwriting analysis performed by questioned document examiners, which encompasses aspects of handwriting analysis beyond the scope of CEDAR-FOX's capabilities. CEDAR-FOX is an interactive software system to assist the document examiner in comparing handwriting samples. Nevertheless, CEDAR-FOX does offer a methodical and rigorous application of well-accepted computational and probability theories to spatial characteristics of handwriting analysis. In so doing, it significantly limits potential user bias in making measurements and comparisons of the handwriting. The theory and operation of CEDAR-FOX have been presented in a published paper. [31] In general, the system's verification accuracy, when presented with at least a half-page of handwriting, has been shown to be approximately 97%. When the amount of writing present is smaller, the accuracy is lower – reaching approximately 90% when only one line of writing is present.

One of the challenges associated with computer-assisted handwriting analysis is conveying a sense of the accuracy or limitations of statistical values in a way that is meaningful to both scientists and courtroom jurors. In 2007, a more refined writer verification experiment was conducted using CEDAR-FOX [32]. As part of the experiment, LLR ranges were quantified and introduced as a nine-point scale similar to that used by questioned document examiners. Although CEDAR-FOX removes many sources of potential human bias through its handwriting analysis features, there remains some level of human interaction in either the design or operation of the system. The potential human bias in the use of CEDAR-FOX, on a relative basis, should be more limited than in traditional handwriting analysis. Further, because the system is subject to ongoing testing, experimentation, and verification, any residual bias may be explored more methodically, either scientifically, or on cross-examination in a judicial proceeding.

A handwriting study was undertaken to complement the 2002 study. [33] The new study, published in 2008, was designed to use an updated version of the CEDAR-FOX system to assist in the analysis of handwriting of a cohort of 206 pairs of twins. The goal of extending the 2002 individuality study to twins involved, among other items, comparing performance of the "automated" analysis of handwriting of twins with the previous analysis of the general population, and (2) determining performance of the CEDAR-FOX system when the textual content of the questioned and known writing is different.

When compared to the 2002 individuality study, the 2008 study made use of additional handwriting features, a different feature extraction method (automatic versus manual), more advanced similarity computations in feature distance space, and a different decision algorithm (a naïve Bayesian classifier). The average error rate for identical twins was 28.4% and the average rate of error for fraternal twins was 11.3%. The average error rates for twins were higher than for non-twins. The average error rate for non-twins remained at approximately 4%, similar to the result in the 2002 study. The 2008 study concluded that "the current system is based on a set of simple features. The use of better features, e.g.., those with a cognitive basis such as the ones used by questioned document examiners, and higher accuracy classification algorithms should further decrease the error rates. As expert human performance has been shown to be significantly better than that of lay persons, many sophisticated improvements are likely to be needed to reach the higher goal [i.e. the level of performance of questioned document examiners]."

8 Summary and Conclusion: The Future Role of Computer-Assisted Handwriting Analysis in Judicial Proceedings

In responding to the assessment of the National Research Council that the scientific basis for handwriting comparison needs to be strengthened, it seems clear that further research will need to be conducted to accomplish this objective. Computer-assisted handwriting analysis has been used in research, business and financial sectors for some time. A number of systems have been available to assist questioned document examiners in performing their work in forensic applications, either by making some tasks simpler, or by assisting in narrowing the number of items for them to review. As experience with systems like CEDAR-FOX grows, benchmarks of performance may be established, leading to the development of more testing, establishment of standards for the systems, and computation of recognized error rates.

Research has not disclosed the level of acceptance of computer-assisted handwriting analysis within the questioned document examiner community. Similarly, research has not yielded an indication of whether there is "general acceptance" in the field in which computer-assisted handwriting analysis belongs (under the Frye standard); nor does it appear that any study has been conducted or published to determine the "degree of acceptance" of computer-assisted handwriting analysis methodology within the "relevant community" (under the Daubert standard). While the issue of general acceptance of the methodology used by questioned document examiners appears to have been settled by appellate courts in the judicial system, there had been vigorous debate over whether the methodology belonged to the community comprised of questioned document examiners, or to a broader group of scholars and scientists. This debate was reflected in the U.S. judicial decisions [34], as well as peer review of the handwriting studies [35]. In determining whether and when computer-assisted handwriting analysis achieves general acceptance, the relevant community will need to be ascertained by the courts, in order to determine the admissibility of the computer-assisted methodology and results.

Daubert teaches that a "reliability assessment does not require, although it does permit, explicit identification of a relevant scientific community and an express determination of a particular degree of acceptance within that community". There also is guidance from a Frye jurisdiction, (New York): "In defining the relevant scientific field, the Court must seek to comply with the Frye objective of obtaining a consensus of the scientific community. If the field is too narrowly defined, the judgment of the scientific community will devolve into the opinion of a few experts. The field must still include scientists who would be expected to be familiar with the particular use of the evidence at issue, however, whether through actual or theoretical research."[36]. The National Research Council, in reviewing the scientific basis for handwriting comparisons, observed that "questions about documents arise … in any matter affected by the integrity of written communications and records". This observation may provide a cornerstone for analysis of the relevant community under either the Frye or Daubert standard. It is expected that, as computer-assisted handwriting analysis becomes more widely used and recognized, and handwriting analysis research continues to advance, the debate over the relevant community to which it belongs will resume.

Because courts, in making rulings in the context of <u>Frye</u> or <u>Daubert</u> hearings, are not required to follow rules of evidence, the admissibility of the computer-assisted methodology, computations, and results themselves have not been the subject of judicial evidentiary rulings. To date, the role of computer-assisted handwriting analysis in the judicial system has been limited to supporting, from a scientific viewpoint, the fundamental premise of the individuality of handwriting, and the ability to test and ascribe potential error rates to handwriting analysis. As computational systems and algorithms become more robust, and the accuracy and reliability of those systems improve, their performance may approach levels identified with the performance of questioned document examiners. As computational systems develop over time, parties in judicial proceedings may decide to offer direct evidence of the newer computational methods, as well as the results of the methods, either in <u>Frye</u> or <u>Daubert</u> hearings, or at trial. The offer of proof may come either as a computation performed by a questioned document examiner, or by a witness properly trained as a user of the computer-assisted handwriting analysis system. It does not appear, to date, that such an effort has been undertaken in a published case. In the event that both traditional questioned document analysis and computer-assisted handwriting analysis are conducted and offered as evidence in the same proceeding, and the <u>Frye</u> and <u>Daubert</u> standards are satisfied, the performance of the questioned document examiner may be measured against the output of the computational analysis. In this manner, the scores produced by the human examiner and the computational system may be compared and contrasted, permitting opportunity either for cross-validation, or for cross-examination.

References

1. National Research Council: Strengthening Forensic Science in the United States: A Path Forward, pp. 4–11. National Academies Press, Washington DC (2009)
2. Wigmore, Evidence §1367 (Chadbourn rev.) (1974)
3. Frye v. United States, 54 App. D.C. 46, 293 F. 1013 (D.C. Cir (1923)
4. Martin, M.M., Capra, D.J., Faust, A., Rossi, F.: New York Evidence Handbook, Novel Scientific Evidence, Sec. 7.2.3., 2nd edn., p. 586. Aspen Publishers, New York City (2003)
5. 509 U.S. 579 (1993)
6. 526 U.S. 137 (1999)
7. National Research Council: Strengthening Forensic Science in the United States: A Path Forward, pp. 5–29. National Academies Press, Washington DC (2009)
8. Farrell, R.: Richardson on Evidence (Cum supp. 1997-2008) Sec. 7-318., 11th edn., p. 485. Brooklyn Law School, Brooklyn (1995)
9. People vs. Silvestri, 44 N.Y. 2d 260, 266 (1978)
10. Nimely vs. City of New York, 414 F. 3d 381, 396 (2nd Cir. 2005)
11. United States vs Williams, 506 F. 3d 151, 162 (2nd Cir. 2007)
12. U.S. v. Starczepyzel, 880 F. Supp. 1027, 1037. (S.D.N.Y. 1995)
13. U.S. v. Saelee, 162 F. Supp. 2d, 1097, 1102. (D. Alaska 2001)
14. U.S. v. Crisp, 324 F. 3d. 261, 280. (4th Cir. 2003)
15. Starczepyzel, 880 F. Supp. at 1037
16. U.S. v. Hines, 55 F. Supp. 2d. 62, 68 (D. Mass 1999)

17. National Research Council: Strengthening Forensic Science in the United States: A Path Forward, pp. 5–30. National Academies Press, Washington DC (2009)
18. Kuckuck, W., Rieger, B., Steinke, K.: Automatic Writer Recognition. In: Proc. 1979. Carnahan Conf. on Crime Countermeasures. University of Kentucky, Lexington (1979)
19. Found, B., Rogers, D., Schmittat, R.: Matrix Analysis: A Technique to Investigate the Spatial Properties of Handwritten Images. J. Forens Doc. Exam (Fall 1998)
20. Plamondon, R., Srihari, S.N.: On-line and off-line handwriting recognition: A comprehensive survey. IEE Transactions on Pattern Analysis and Machine Intelligence 22(1), 63–84 (2000)
21. Srihari, S.N., Leedham, C.G.: A survey of Computer Methods in Forensic Document Examination. In: Proceedings of the 11th Conference of the International Graphonomics Society, November 2-5, 2003. IGS, Scottsdale (2003)
22. Srihari, S.N., Cha, S.H., Arora, H., Lee, S.: Individuality of Handwriting. J. Forens Sci. 44(4), 856–872 (2002)
23. Huber, R.A., Headrick, A.M.: Handwriting Identification: Facts and Fundamentals. CRC Press, Boca Raton (1999)
24. Fed. R. Evid. 104 (a)
25. Janopoulos v Harvey L. Walner & Assoc., 866 F. Supp. 1086 (N. D. Ill 1994)
26. U.S. v. Prime, 431 F. 3d 1147, 1153 (9th Cir. 2005)
27. U.S. v. Yagman 2007 WL 4409618 (C.D.Cal. 2007)
28. U.S. v. Gricco, 2002 WL 746037, *3-*4
29. Srihari, S.N., Zhang, C., Tomai, S.J., Lee, Z., Shi, Y.C., Shin, A.: A system for handwriting matching and recognition. In: Symposium on Document Image Understanding Technology (SDIUT 2003), Greenbelt, MD (2003)
30. Version 1.3, March 3, 2008. 2003-2008 The Research Foundation of State University of New York. All Rights Reserved 2003-2008 CEDAR Tech, Inc. All Rights Reserved, Portions of the product were created using LEADTOOLS 1991-2002 LEAD Technologies, Inc. (2003-2008)
31. Srihari, S., Srinivasan, H., Desai, K.: Questioned Document Examination Using CEDAR-FOX (2007)
32. Kabra, S., Srinivasan, H., Huang, C., Srihari, S.: On Computing Strength of Evidence for Writer Verification. In: ICDAR (2007)
33. Srihari, S.N., Huang, C., Srinivasan, H.: On the Discriminability of the Handwriting of Twins. J. Forens. Sci. 53(2), 430–446 (2008)
34. U.S. v. Starczecpyzel, 880 F. Supp 1027, 1039 (1995)
35. Sax, M.J., Vanderhaar, H.: On the General Acceptance of Handwriting Identification Principles. J. Forens Sci. 50(1), 119–124 (2005); Kelly, J.S., Carney, B.B., Sax, M.J., Vanderhaar, H.: On the General Acceptance of Handwriting Principles. J. Forens Sci. 50(1), 119–124 (2005); J. Forens. Sci. 50(5) (September 2005); Purdy, D.C., Sax, M.J., Vanderhaar, H.: On the General Acceptance of Handwriting Identification Principles. J. Forens Sci. 50(1), 119–126 (2005); Author's response; J. Forens Sci. 5(5) (2005)
36. People v. Wesley, 83 N.Y. 2d 417, 438 (citing Giannelli, The Admissibility of Novel Scientific Evidence: Frye v. United States, a Half-Century Later 80) Columbia Law Review (1197, 1209 – 1210)

Analysis of Authentic Signatures and Forgeries

Katrin Franke

Norwegian Information Security Laboratory, Gjøvik University College, Norway
kyfranke@ieee.org

Abstract. The paper presents empirical studies of kinematic and kinetic signature characteristics. In contrast to previous studies a more
in-depth analysis is performed which reveals insides on differences and
similarities of authentic and mimicked signing movements. It is shown
that the signing behavior of genuine writers and impostors is only likely
to differ in terms of local characteristics. Global characteristics can easily
be imitated by skilled forgers. Moreover, it is shown that authentic writing characteristics cover a broad value range that might interfere with
value ranges of unsophisticated forgeries. In our experiments signing behavior of 55 authentic writers and of 32 writers mimicking signature
samples with three different levels of graphical complexity is studied. We
discuss implications for ink-trace characteristics on paper and provide
recommendations for implementing computer-based analysis methods.

Keywords: signature verification, handwriting behavior, forensic examination, combined on-/offline analysis.

1 Introduction

Handwriting movements, including their planning and control, have been studied
quite intensively and from various perspectives, e.g. with regard to psychomotor, medication or brain-activity aspects. However, it is always worthwhile to
critical review and *cross-validate* those previous studies, in particular when they
have been conducted in other domains with different research questions, experimental setups and computational approaches. As computer scientist, we perform the study presented here in line with our research on new methods for the
computer-based forensic analysis of signatures [1]. Currently, we concentrate on
biomechanical handwriting aspects, especially on those movement characteristics
that can affect the ink deposit on paper.

According to psychomotor theory, forging another person's handwriting or
signature is an untrained motor task that demands feedback, e.g. [2,3,4]. Especially visual feedback is needed in order to simulate general shape characteristics
in the best possible way. Since it is not possible to resort to previously memo-
rized letter shapes, the planning / execution of the handwriting movement has to
be performed step by step [5]. As a result it is to be assumed that forgers reduce
their writing velocity, pause frequently, increase limb stiffness and apply higher
pen force, e.g. [2,6,7,8,9,10,11,12,13,14,15]. Researchers from the psychomotor

Z.J.M.H. Geradts, K.Y. Franke, and C.J. Veenman (Eds.): IWCF 2009, LNCS 5718, pp. 150–164, 2009.

field [12,16,17] postulate also a difference in the *power spectral density (PSD)* of the writing-velocity signals for authentic and mimicked handwriting.

The objective of our new, independent study is threefold: (1) We wanted to replicate previous experiments and gain more knowledge on the distribution of a particular writing characteristic; for authentic and mimicked signing behaviors. (2) While studying the literature we observed contradicting statements on pen force characteristics that needed to be solved, and (3) We realized that in the discussions of writing kinetics the transition of movements in the air towards movements on the paper / writing support have not been addressed so far. In the following we motivate these three objectives for our research in more detail.

Add. 1) Pen displacement: velocity, pauses, and variation: Replicating an experiment conducted previously is always recommended in order to obtain confidence in the results reported. This holds in particular, when the data sets used were rather small or the computational procedure used by others raised questions. In our study we wanted to observe differences in authentic and mimicked handwriting on our own independently recorded writings. In particular, we aimed at a forensic-like data set with a large number of skilled forgeries. Next to that we questioned that all forgeries will write less fluent, since we observed forgers writing fluently and producing a high correspondence in the signature shape. So, we aimed not only to look at general / average writing characteristics, but also to study and to report parameter distributions in great detail. We were also curious to determine writing pauses, which are also referred to as stop points, since writers seem to hold the pen still at those points. Is this only done in case of mimicking another persons signature, or show genuine writers similar behavior? Last but not least we were questioning the computational procedure used in the PDS-study, which is not sufficiently described in these papers.

Add. 2) Pen force: In signature analysis, be it computer-based or in the classical forensic domain, the applied pen-tip force is considered as an important source of information, e.g. [2,6,12,18,19,20]. Several studies have been conducted in order to determine the discriminative value between genuine and mimicked handwriting. To this end, forensic practitioners mainly focus on trace thickness, ink intensity and pen groove, e.g. [6,18,19,21]. In contrast, the researchers from human movement science use modern, electronic devices to capture applied pen-tip forces during the writing process, e.g. [2,12,13,14]. While preparing our studies we discovered one contradicting statement between forensic and psychomotor experts that seemed worthwhile to investigate in more detail. Previous works in human movement science report an increase of pen-tip force in the process of mimicking another person's handwriting, whereas forensic literature only refers to a variation in the rhythm of applied pen pressure. The difference is of great interest, since a permanently increased pen-tip force would lead to more ink deposit on paper. The phenomenon would affect the ink-trace pattern as a whole - or in other words - it would constitute a change of a *global* characteristic, while the other case only implies a local variation / divergence. In addition, it needs to be examined, whether genuine writers exist who apply a larger amount of pen-tip force; people who have possibly less writing practice / skills or people performing heavy manual work.

Add. 3)   Pen landing and lift: Besides the applied pen force, we were curious to find out how the transition from pen-tip trajectories in the air to movements in the xy-writing plane, where the pen is pressed onto the paper, is performed. The most interesting question in this context was, whether there are genuine writers that have longer delays / reaction times before beginning to move the pen. What happens during this time? Do the writers keep the pen still or do they already exert force on the pen tip? Recent studies [22,23,24] have also revealed that ink traces show distinct ink-deposit characteristics at the beginning and at the end, which can be used for writer identification. However, one has to keep in mind that effects of the writing material used are superimposed [25]. A systematical analysis of pen landings and lifts of genuine writers will also show whether authentic writers produce certain ink-trace characteristics, in particular thick stumpy stroke beginnings and endings, that could be confused with the characteristics of forged handwriting.

Our performed studies of authentic and forged signing behaviors are described in the following two sections. Section 2 is covering details of the applied analysis methods, signature data and procedures; Section 3 is presenting and discussing the results. More specific, the kinematics of the writing process as velocity, pauses and power spectral density are discussed in Section 3.1. Kinetics, the change of applied pen-tip force as well as the transition from the aerial pen movement to pen force, are the objects of the discussion detailed in Section 3.2. Finally, Section 4 concludes the discussions and provides recommendations for computational method development.

2   Method

2.1   Signature Data

The NISDCC signature collection [26] has been acquired in the framework of the WANDA project [27], and became public available with the 2009 ICDAR Signature Competition [28]. The collection contains simultaneously recorded online and offline samples: authentic signatures from 12 writers (5 authentic signatures per writer), and forged signatures from 32 writers (5 forgeries per authentic signature). Writers were students in Psychology and Artificial Intelligence as well as employees of the University of Njimegen. Among these writers were five experts in handwriting recognition and signature verification. In addition, we used signatures of 55 authentic writers, each writing their genuine signature 30 times. Writers were students and staff of the Fraunhofer IPK, Berlin. This FhG-IPK signature collection is not public available.

2.2   Procedure

From the NISDCC-online data we selected $3 \times 32 \times 5$ forgeries. The forgeries are for *three different* authentic signatures. These signatures are of low (L), medium

(M) and high (H) *graphical complexity* [7,29,30,31]. Therefore, the samples require three different levels of *graphical skills* [32] on the side of the 32 impostors.

Writers were instructed to practice in writing authentic or copying forged signatures as much as they found necessary. Forgers were requested to copy signatures as fluently as possible, while trying to mimic the shape of the authentic signature as much as possible.

2.3   Materials

The NISDCC-online and FhG-IPK data was recorded using: Wacom Intuos2 A4-oversized tablet (XD-1212-U) [33]; Wacom Intuos inking pen (XP-110-MOD); Windows98, 1st Edition (Release 4.10, 1998); Wacom tablet driver Version: 4.75-9; and the *SIC Tablet* tablet-interface developed by Fraunhofer IPK, Berlin.

It must be noted that the writing tablet used requires a minimal pen-tip force of 0.3 N [33] in order to activate force-data capturing. If the pen-tip force is less, the tablet gives "0" as the measured value. This effect can be explained by the measurement principle, which in the case of this particular tablet is a capacitive one. Nevertheless, in order to obtain an overview of the biomechanical writing characteristics, the available tablet is considered as suitable.

2.4   Analysis

In our study we concentrate on the analysis of biomechanical signing characteristics. In accordance with previous studies, e.g. [2,11,12,13,14,15], the following kinematic and kinetic parameters have been selected: *pen displacement* including *writing velocity, pauses, power spectral density* and *pen-tip force, landing and lift.* Pen-orientation was addressed in our previous study [34].

In order to obtain a general overview of the biomechanical characteristics of authentic and forged handwriting, we decided to use descriptive statistics. From our perspective, it is most important to have an in-depth understanding of the data first, before one can aim at implementing sophisticated computer-based analysis methods. In the next sections, figures will be reported that contain a considerable amount of detail. The rationale for this choice of visualisation method is that this study concerns the actual details of individual differences. Box plots give both mean and variance information of a data distribution and are therefore preferred in the forensic domain, where individuals are usually not averaged away, since the discrimination between them is the goal. We opted for the median of a distribution, since it is more robust than the mean / average value, in particular if the data distribution is non-symmetrical (skewed). In addition, we report the min, max, quantile 0.3 and quantile 0.7 values of a distribution since it is especially useful to get an impression of the data.

Analysis of Writing velocity: To validate movement characteristics of authentic and mimicked singing behavior we computed the writing velocities along the tangential direction of the trajectories, and visualize absolute writing-velocity distributions in box plots as explained before.

Analysis of Writing pauses - stop / break points: In the data analysis we defined pauses according to recommendations by our colleagues from Nijmegen [16]. Accordingly, stop points are such coordinate sequences that have a minimum tangential velocity of less than 5 % of the average writing velocity.

Analysis of Variance density spectrum: Researchers from the psychomotor field [12,16,17] postulate a difference in the *power spectral density (PSD)* of the velocity signals (x, y, tangential writing direction) for authentic and mimicked handwriting. We would like to review some fundamentals of signal processing in order to lay a solid fundament for further discussions. The (PSD) is commonly used in (electrical) signal analysis and often applied to determine how the power (or variance) of a time series is frequently distributed. Mathematically, the PSD is defined as the Fourier Transformation of the autocorrelation sequence of the time series [35]. An equivalent definition of PSD is the squared modulus of the Fourier transformation of the time series, scaled by a proper constant term. Nevertheless, three primary aspects certainly have to be taken into consideration for the numerical computation of the PSD [36]:

1. *Frequency aliasing* - The time signals sampling period has to be chosen to ensure that frequency aliasing due to time sampling is negligible.
2. *Frequency resolution* - is determined by the number of sampling points; the higher the number, the higher the frequency resolution. In order to increase the basic resolution of the *Fast Fourier Transformation* the length of the time series can be artificially extended by adding zero value samples. However, this will lead to smoothing in the frequency domain.
3. *Effects of truncation* - If the time series is of infinite duration, it needs to be truncated to a finite length. This will lead to leakage, ripples and create accuracy problems.

The implications for the handwriting signal, and for forged handwriting in particular, are as follows: The sampling period is determined by the electronic pen-tablet and can be, for example, 5 ms as in our cases. Since forged handwriting is mainly written at a much slower pace, more pen positions need to be / are captured in order to sense the complete (signature) trajectory. As a rule of thumb factor 3-4 can be mentioned for the increase of recorded pen positions. In order to avoid truncation effects in the numerical computation of the PSD, the length of the forgery time series needs to be taken into consideration. If one aims at a comparable frequency resolution, genuine and forged specimens have to be filtered on the basis of the same number of sample points, or appropriately post-processed. Therefore, the genuine time series needs to be extended to the length of the forgery time series, e.g. by means of *zero padding*.

Analysis of Pen-tip force, pen landings and lift: For data analysis we opted for descriptive statistics using box plots. We also choose to display some real data exemplarily.

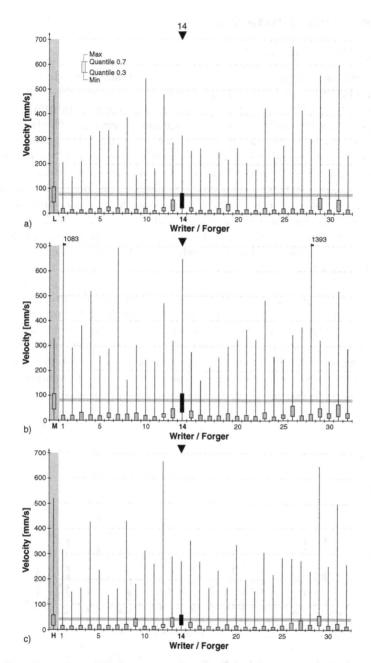

Fig. 1. Box plot of tangential writing velocity in mm/s. The 32 impostors faked three genuine signatures classified as being of (a) low, (b) medium, and (c) high graphical complexity. The velocities are represented by the smallest and largest observation as well as by their inter-quantile range. The entry on the far left shows the velocity range of the authentic writer. Note that *forger 14* writes at a comparable speed.

3 Results and Discussion

3.1 Pen Displacement

Writing velocity of forged signatures: To validate movement characteristics of forged handwriting, signatures of three genuine writers and those of 32 forgers simulating each of the authentic are considered. Figure 1 illustrates absolute writing velocities (velocities along the tangential direction of the trajectories). The three diagrams show velocities for each of the authentic signature samples separately. The left-hand row contains values of the genuine writer, whereas the others represent data of the forgeries. The obtained data (Figure 1) clearly reveals that the movement velocity of the majority of forgers is significantly lower than that of authentic writers, if one looks at the inter-quantile ranges only. However, impostors can reach top velocities that are much higher than those of the genuine writers. Moreover, some forgers are able to simulate general velocity characteristics well. *Forger 14*, who seems to have excellent motor skills, is a good example.

Writing pauses in forged signatures: Figure 2 illustrates the writing pauses for the three authentic sample signatures as well as for the fakes produced by the 32 impostors. Surprisingly, even genuine writers occasionally pause during writing, as can be seen in the case of *writer H*, who produces the most graphically complex signature. In addition, it is clearly recognizable that the majority of impostors pause repeatedly, obviously to program the next movement sequence.

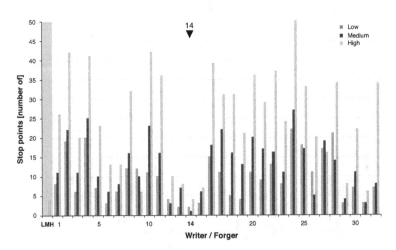

Fig. 2. Bar chart of observed pen stops. Leftmost bars, labeled *LMH*, contain data of three writers *L*, *M* and *H*, who provided their authentic signatures of supposedly low, medium and high graphical complexity. The other bars show the pen stops of 32 impostors, simulating each one of the authentic signature specimens. Once again, *forger 14* is able to imitate another person's signature very well, and even with very few hesitations (see also Figure 1).

The graphical complexity of the signature pattern to be forged significantly influences the number of breaks. *Impostor 14*, once again, emulates handwriting both fluently and with very few hesitations. This is an excellent example of the fact that forgers, depending on their graphical and motor skills, are indeed able to simulate another person's handwriting.

Writing velocity of genuine signatures: Following a similar procedure as before we generated graphical plots (Figure 3). The velocity characteristics of the majority of the 55 authentic writers are rather unspectacular, with inter-quantile ranges between 100 mm/s and 200 mm/s. The top speeds are between 400 and 600 mm/s. However, the movements of *writer 1* are extremely fast. The inter-quantile range in his case is between 380 mm/s and 570 mm/s; his top speed is 745 mm/s. These data, in particular, have been carefully cross-checked and confirmed by similar results.

Variance density spectrum of velocity signals: In view of all the requirements outlined in Section 2.4, the reported results [16] could not be replicated, although the same test data were used. We observed that activations in the higher frequency band, i.e. between 30 - 40 Hz, are not stable for all impostors, and not even for the five probes produced by one single forger.

3.2   Pen Force, Landing and Lift

In order to obtain more insights, we examined the captured pen-tip forces of 55 genuine writers and 32 impostors. For data analysis we opted for descriptive statistics as before. Figures 4 and 5 provide the data of the 55 genuine writers and the 32 forgers, respectively.

Pen forces of genuine signatures: The inter-quantile ranges are quite differently distributed. Hence, it cannot be assumed that a genuine writer would never apply a larger amount of pen force. *Writer 31* and *39* are good examples. In addition, it is recognizable that almost any pen-force range is possible. Rather narrow inter-quantile ranges attest to the fact that the applied force of individual writers does not vary to any great extent. In this case it is also irrelevant whether writers prefer more or less force.

For example, *writer 27* and *31* apply a lot of pen force, while *writer 21* and *34* virtually seem to glide across the writing paper. On the other hand, there are writers with extensive inter-quantile ranges, as in the case of *writer 25* and *41*. These writers show a wide spectrum of pen-tip forces. Thus, one can conclude that pen-force signals are indeed individual characteristics that do not follow general rules. Schomaker et al. [2] have investigated this circumstance more systematically. The study revealed that for the majority of writers applied pen-tip forces are uncoupled controlled writing-process parameters that do not correlate with the pen displacement. Unfortunately, forged handwriting was not addressed in this investigation.

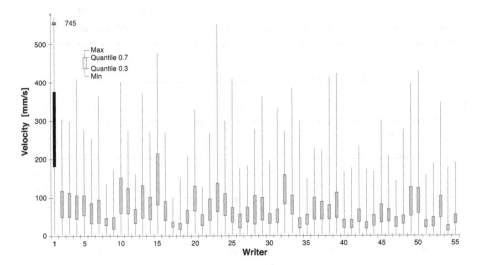

Fig. 3. Absolute writing velocity in mm/s for 55 writers producing their genuine signature 30 times. The observed velocities are represented by their smallest and largest observation as well as by their inter-quantile range. Left-most writer is an outlier.

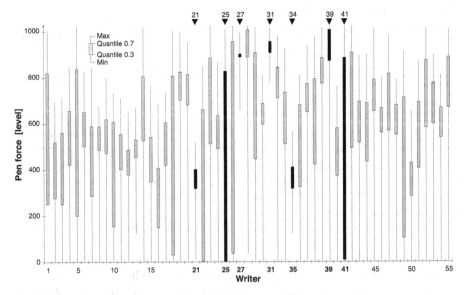

Fig. 4. Applied pen forces in *level* for 55 writers producing their genuine signature 30 times. The pen-force information is represented by its smallest and largest observation as well as by its inter-quantile range.

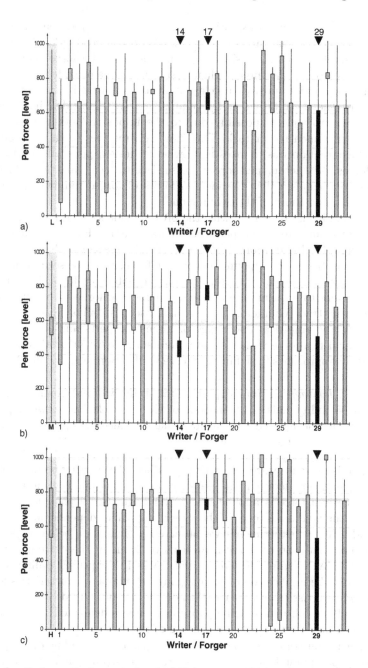

Fig. 5. Applied pen-tip forces in *level* of 32 impostors faking three genuine signatures that can be classified as being of (a) low, (b) medium and (c) high graphical complexity. The pen-force information is represented by the smallest and largest observation as well as by its inter-quantile range. The pen-tip force of the authentic writer is shown on the far left. For detailed discussions the reader is referred to page 160, left.

Pen forces of forged signatures: Figure 5 illustrates captured pen-force data of three genuine writers that are indexed as *L*, *M* and *H* according to the supposed graphical complexity of their signatures. In addition, the force ranges of the 32 forgers who copied the three authentic specimens are provided.

The given data plots provide no clear indication of forged handwriting when compared to the data provided by genuine writers. Extended inter-quantile ranges are more frequent, possibly due to the impostors' permanent adaptation to the untrained motor task. This, however, requires further systematical studies. Our intention was to investigate whether forgers generally apply higher pen-tip forces, or whether pen forces are subject to local variations. From the plots 5 a-c one can see that the median forces of forged handwriting are *not* always higher than those of the genuine one.

For the time being we remain with the proven evidence that some forgers are indeed able to simulate global pen-force characteristics. Forgers 14 and 17 are only two examples. In addition, it has to be taken into account that some forgers may follow the strategy of applying *less* pen-tip force, as can be seen in the case of forger 29. Therefore, we conclude that median or global pen-force characteristics are not sufficient to discriminate between genuine and forged handwriting, and that global ink-trace characteristics are not useful in an analysis procedure. According to the empirical knowledge of forensic experts and some computer scientists, local pen-force variations are more suitable for an analysis, which is supported by approaches in online signature verification, e.g. [37,38,39].

Pen lifts and landings: In our experimental study we used genuine signature samples of 55 individuals. The analysis of the interrelation between pen displacements and the applied pen-tip force has revealed two primary strategies that are followed by the genuine writer:

1. The pen is moved towards the writing surface, the starting point for the ink trace is somehow focused, the pen-tip force is continuously increased, and once a certain force is applied, the writer begins to displace the pen. The delay of the pen movement (in the xy-writing plane) depends on the writer's attitude. In our experiments we observed a time shift between pen-point kinematics and pen-tip force from 5 ms to 20 ms in the extreme. In order to provide an example Figure 6a illustrates the captured online data for *writer 12*. The implication of such signing-behavioral characteristics for ink-trace characteristics on paper is that due to the prolonged contact between the tip of the ink pen and the paper more ink is deposited, and hence ink drops can be observed at the beginning of the stroke.

2. The other group of writers performs a soft (plane-like) landing of the pen tip, whereby pen-point kinematics are continuously transferred into pen force. A delay between pen force and pen displacement was not observable. In order to provide an example Figure 6b shows the interrelation between pen displacement and pen force for *writer 7*. Note that in cases of high writing velocity the exact moment when the pen touches the writing surface may not be recorded. This is due to the time-discrete sampling of the signal.

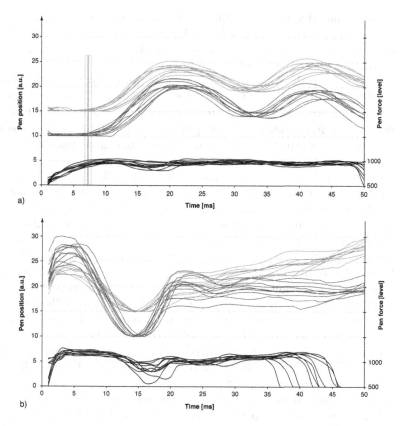

Fig. 6. Examples of primary strategies for pen landing in authentic handwriting: (a) The pen is moved towards the writing surface. Then, while holding the pen still the pen-tip force is increased, and after a pause the pen displacement is launched. (b) The pen trajectory in the air is directly converted into pen movements on the writing surface with specific pen-tip forces (soft-landing).

Consequently, captured pen-force data can comprise higher initial values, e.g. 500 pressure levels. Nevertheless, ink-trace effects caused by the undelay writing strategy are soft transitions in the ink deposition. Depending on the fluidity of the ink there can be an increase in ink intensity [25], e.g. for viscous and solid inks, or merely an increase in stroke width, e.g. for fluid inks.

4   Conclusions

The experiments described in this paper dealed with biomechanical handwriting characteristics, which are involved in the production of authentic and forged signatures. The objectives of this study were (1) to provide sufficient details on the kinematics and kinetics during the forged and genuine act of signing, (2) to solve conflicting statements on pen force characteristics, and (3) to investigate the transition of pen movements in the air towards writing movements on the paper / writing support (pen landing and lift.)

In order to study the kinematic and kinetic characteristics of the signing behavior (Section 3.1 and 3.2) the writing movements of 55 writers producing their genuine signature and of 32 writers forging three sample signatures were recorded. Three different signatures were provided for forging, i.e. of low, medium and high graphical complexity. These samples demand three different levels of graphical skills on the side of the impostor.

The analysis of data reveals that the movements of the majority of forgers are significantly slower than those of authentic writers. However, impostors can reach top velocities exceeding those of genuine writers, and certain forgers are able to simulate general velocity characteristics extremely well. It is also clearly recognizable that the majority of impostors perform multiple pen stops. The graphical complexity of the signature pattern to be forged significantly influences the number of breaks. Again, there are impostors, who perform mimicked handwriting not only fluently, but also with very few hesitations. This clearly confirms that forgers, depending on their graphical and motor skills, are indeed able to simulate another person's movements. Global pen-force characteristics do not provide any clear indications regarding the authenticity of a questioned handwriting specimen. It has been proved that forgers do not generally apply higher pen-tip forces. It has also been revealed that some forgers follow the strategy of applying less pen-tip force.

The only possible conclusion is that *variations* in the movement time series are reliable with regard to writer-specific characteristics. Median or global characteristics are not sufficient to distinguish between genuine and forged handwriting. Consequently, global ink-trace characteristics are unsuitable for an analysis procedure. Only the local, inner ink-trace characteristics as well as variations in ink intensity and line quality can provide reliable information in the forensic analysis of signatures.

References

1. Franke, K.: The Influence of Physical and Biomechanical Processes on the Ink Trace - Methodological foundations for the forensic analysis of signatures. PhD thesis, Artifical Instelligence Institute, University of Groningen, The Netherlands (2005)
2. Schomaker, L., Plamondon, R.: The relation between pen force and pen point kinematics in handwriting. Biological Cybernetics 63, 277–289 (1990)
3. van Galen, G.: Handwriting: Issues for a psychomotor theory. Human Movement Science 10 (1991)
4. Smyth, M., Silvers, G.: Functions of vision in the control of handwriting. Acta Psychologica 65, 47–64 (1987)
5. van Galen, G., van Gemmert, A.: Kinematic and dynamic features of forging another person's handwriting. Journal of Forensic Document Examination 9, 1–25 (1996)

6. Conrad, W.: Empirische Untersuchungen zur Differentialdiagnose zwischen verschiedenen Unterschriftsgattungen. Zeitschrift für Menschenkunde 35, 195–222 (1971) (in German)
7. Found, B.: The Forensic Analysis of Behavioural Artefacts: Investigations of Theoretical and Analytical Approaches to Handwriting Identification. PhD thesis, LaTrobe University Bundoora (1997)
8. Halder-Sinn, P., Funsch, K.: Die Diagnose von Haltepunkten - Mehr als ein blosses Ratespiel? Mannheimer Hefte für Schriftenvergleichung 22(3), 3–12 (1996) (in German)
9. Leung, S., Cheng, Y., Fung, H., Poon, N.: Forgery I: Simulation. Journal of Forensic Sciences 38(2), 402–412 (1993)
10. Leung, S., Cheng, Y., Fung, H., Poon, N.: Forgery II: Tracing. Journal of Forensic Sciences 38(2), 413–424 (1993)
11. Thomassen, A., van Galen, G.: Temporal features of handwriting: Challenges for forensic analysis. In: Proc. 5th European Conference for Police and Handwriting Experts, The Hague, The Netherlands (1996)
12. van Gemmert, A., van Galen, G., Hardy, H.: Dynamical features of disguised handwriting. In: Proc. 5th European Conference for Police and Handwriting Experts, The Hague, The Netherlands (1996)
13. van Gemmert, A., van Galen, G.: Dynamic features of mimicking another person's writing and signature. In: Simner, M., Leedham, C., Thomassen, A. (eds.) Handwriting and drawing research: Basic and applied issues, pp. 459–471. IOS Press, Amsterdam (1996)
14. van den Heuvel, C., van Galen, G., Teulings, H., van Gemmert, A.: Axial pen force increases with processing demands in handwriting. Acta Psychologica 100, 145–159 (1998)
15. van Gemmert, A.: The effects of mental load and stress on the dynamics of fine motor tasks. PhD thesis, Catholic University, Nijmegen, The Netherlands (1997)
16. Beenders, M.: On-line and off-line signature verification techniques from a psychomotor perspective. Master's thesis, Catholic University, Nijmegen, The Netherlands (2003)
17. van Galen, G., van Doorn, R., Schomaker, L.: Effects of motor programming on the power spectral density function of finger and wrist movements. Journal of Experimental Psychology: Human Perception and Performance 16, 755–765 (1990)
18. Baier, P.: Schreibdruckmessung in Schriftpsychologie und Schriftvergleichung - Entwicklung und experimentelle Überprüfung neuer Registrierungsverfahren (1980) (in German)
19. Deinet, W., Linke, M., Rieger, B.: Analyse der Schreibdynamik. Technical report, Bundeskriminalamt Wiesbaden, Thaerstraße, Wiesbaden, Germany, vol. 11, p. 65193 (1987) (in German)
20. Plamondon, R., Lorette, G.: Automatic signature verification and writer identification - the state of the art. Pattern Recognition 22, 107–131 (1989)
21. Maus, E.: Schriftdruckmessung: Grundlagen, Methoden, Instrumente. Sciptura, Lothar Michel Weinheim Bergstraße (1996) (in German)
22. Doermann, D., Rosenfeld, A.: Recovery of temporal information from static images of handwriting. In: Proc. of Computer Vision and Pattern Recognition, pp. 162–168 (1992)
23. Doermann, D., Rosenfeld, A.: The interpretation and recognition of interfering strokes. In: Proc. International Workshop on Frontiers in Handwriting Recognition (IWFHR), Tajon, Korea, pp. 41–50 (1993)

24. Schomaker, L., Bulacu, M., Erp, M.: Sparse-parametric writer identification using heterogeneous feature groups. In: Proc. IEEE International Conference on Image Processing (ICIP), Barcelona, Spain, pp. 545–548 (2003)
25. Franke, K., Rose, S.: Ink-deposition model: The relation of writing and ink deposition processes. In: Proc. 9th International Workshop on Frontiers in Handwriting Recognition (IWFHR), Tokyo, Japan, pp. 173–178 (2004)
26. Blankers, V.L., van den Heuvel, C.E., Franke, K., Vuurpijl, L.: The ICDAR 2009 signature verification competition with On- and Offline Skilled Forgeries (2009), http://sigcomp09.arsforenscia.org
27. Franke, K., Schomaker, L., Veenhuis, C., Taubenheim, C., Guyon, I., Vuurpijl, L., van Erp, M., Zwarts, G.: WANDA: A generic framework applied in forensic handwriting analysis and writer identification. In: Abraham, A., Köppen, M., Franke, K. (eds.) Design and Application of Hybrid Intelligent Systems, Proc. 3rd International Conference on Hybrid Intelligent Systems (HIS 2003), pp. 927–938. IOS Press, Amsterdam (2003)
28. Blankers, V.L., van den Heuvel, C.E., Franke, K., Vuurpijl, L.: The icdar 2009 signature verification competition (sigcomp 2009). In: Proc. 10th International Conference on Document Analysis and Recognition, (ICDAR 2009) (2009) (in press)
29. Alewijnse, L.C., van den Heuvel, C.E., Stoel, R., Franke, K.: Analysis of signature complexity. In: Proc. 14th Conference of the International Graphonomics Society (IGS 2009) (2009) (in press)
30. Brault, J., Plamondon, R.: A complexity measure of handwritten curves: Modeling of dynamic signature forgery. IEEE Transactions on Systems, Man, and Cybernetics 23, 400–413 (1993)
31. Hecker, M.: Forensische Handschriftenuntersuchung. Kriminalistik Verlag (1993) (in German)
32. Michel, L.: Gerichtliche Schriftvergleichung. De Gruyter (1982) (in German)
33. Wacom Co. Ltd.: (2003), http://www.wacom.com/
34. Franke, K., Schomaker, L.: Pen orientation characteristics of on-line handwritten signatures. In: Teulings, H., van Gemmert, A. (eds.) Proc. 11th Conference of the International Graphonomics Society (IGS), Scottsdale, Arizona, USA, pp. 224–227 (2003)
35. Shiavi, R.: Introduction to Applied Statistical Signal Analysis. Academic Press, San Diego (1999)
36. Chen, C.: Digital Signal Processing - Spectral Computation and Filter Design. Oxford University Press, Oxford (2001)
37. Dolfing, H.: Handwriting Recognition and Verification: A Hidden Markov Approach. PhD thesis, Eindhoven University of Technology, Eindhoven, The Netherlands (1998)
38. Schmidt, C.: On-line Unterschriftenanalyse zur Benutzerverifikation. PhD thesis, RWTH Aachen University (1998) (in German)
39. Wirtz, B.: Segmentorientierte Analyse und nichtlineare Auswertung für die dynamische Unterschriftsverifikation. PhD thesis, Technische Universität München (1998) (in German)

Automatic Line Orientation Measurement for Questioned Document Examination

Joost van Beusekom[1], Faisal Shafait[2], and Thomas Breuel[1,2]

[1] Technical University of Kaiserslautern, Kaiserslautern, Germany
[2] German Research Center for Artificial Intelligence (DFKI),
Kaiserslautern, Germany
joost@iupr.dfki.de, {faisal.shafait,tmb}@dfki.uni-kl.de
http://www.iupr.org

Abstract. In questioned document examination many different problems arise: documents can be forged or altered, signatures can be counterfeited, etc. When experts attempt to identify such forgeries manually, they use among others line orientation as a feature. This paper describes an automatic mean for measuring the line justification and helping the specialist to find suspicious lines. The goal is to use this method as one of several screening tools for scanning large document collections for the potential presence of forgeries. This method extracts the text-lines, measures their orientation angle and decides the validity of these measured angles based on previously trained parameters.

1 Introduction

Questioned document examination is a broad field with many different problems. The questions may be about the age of a document, about the originality of a document [1], about the author of a handwritten document, or about the authenticity of the content of the document. In this context, a reoccurring question is whether the content of a document has been altered or not e.g. by adding additional text to the document.

In the case of a contract, one of the contracting parties may add something to an already signed contract, e.g. by printing an additional lines on the page. Forgery could also be done by pasting printed lines on existing parts of the document and then copying the document in order to obtain an original looking document. In either of these cases, even if it is well done, the chance of having small differences in line orientation is high. The differences may be so small that they cannot be detected by a human eye.

In this paper we present a method to help the examiner to detect these differences more rapidly by automatically detecting misaligned lines in printed text. Tools for doing this manually do exist and seem to be used by the community [2]. Our approach uses an automatic line finder to extract the lines from the documents. This is used to estimate the variance of line orientations on training documents. Using these parameters, the questioned document can be analyzed

Z.J.M.H. Geradts, K.Y. Franke, and C.J. Veenman (Eds.): IWCF 2009, LNCS 5718, pp. 165–173, 2009.
© Springer-Verlag Berlin Heidelberg 2009

for suspicious lines. To our best knowledge this is the first work into the direction of automating this process.

The remaining parts of this paper are organized as follows: Section 2 describes the general approach. Section 3 presents more details about the automatic line finder algorithm used. In Section 4 preliminary results are presented. Section 5 concludes the paper.

2   Description of the Approach

Detecting variations in the horizontal alignment of lines can be done in different ways. One possibility would be to measure one line manually and to define a threshold. This is cumbersome and not very robust. Therefore we follow a trainable and statistically motivated approach: in a first step, the rotation angles of the lines are modelled using a normal distribution. In the second step the questioned document is checked whether the lines fit the model or not.

The alignment of the text-lines is measured by the rotation angle θ of the text-line. For a given a set of training images, the text-lines and their corresponding rotation angle are extracted. As scanned document images tend to vary slightly in the rotation angle, the mean rotation for each page is computed. For all the text-line rotation angles in a page this mean is subtracted to achieve a normal distribution with $\mu_\theta = 0$. Then the standard deviation σ_θ is computed using maximum likelihood estimation.

The obtained parameters are used in the second step to evaluate the alignments of text-lines in a questioned document. Therefore the text-lines are extracted in the same way as for the training. The distribution of the obtained line rotation angles is translated to obtain a mean values of 0. Then for each line angle it is checked whether it is in the 68%, 95% or 99.7% confidence interval or not. Finally this information is added to the image by coloring the lines respectively to obtain a graphical representation of the results.

3   Line Finding

The presented method for forgery detection uses text-line extraction for Roman script text-lines. Although several methods exist for extracting text-lines from scanned documents [3], we use the text-line detection approach by Breuel [4] since it accurately models the orientation of each line [5]. We will first illustrate the geometric text-line model since it is crucial for the understanding of this work.

Breuel proposed a parameterized model for a text-line with parameters (r, θ, d), where r is the distance of the baseline from the origin, θ is the angle of the baseline from the horizontal axis, and d is the distance of the line of descenders from the baseline. This model is illustrated in Figure 1. The advantage of explicitly modeling the line of descenders is that it removes the ambiguities in baseline detection caused by the presence of descenders.

Fig. 1. An illustration of Roman script text-line model proposed by Breuel [4]. The baseline is modeled as a straight line with parameters (r, θ), and the descender line is modeled as a line parallel to the baseline at a distance d below the baseline.

Based on this geometric model of Roman script text-lines, we use geometric matching to extract text-lines from scanned documents as in [4]. A quality function is defined which gives the quality of matching the text-line model to a given set of points. The goal is to find a collection of parameters (r, θ, d) for each text-line in the document image that maximizes the number of bounding boxes matching the model and that minimizes the distance of each reference point from the baseline in a robust least square sense. The RAST algorithm [6,7] is used to find the parameters of all text-lines in a document image. The algorithm is run in a greedy fashion such that it returns text-lines in decreasing order of quality.

Consider a set of reference points $\{x_1, x_2, \cdots, x_n\}$ obtained by taking the middle of the bottom line of the bounding boxes of the connected components in a document image. The goal of text-line detection is to find the maximizing set of parameters $\vartheta = (r, \theta, d)$ with respect to the reference points $\{x_1, x_2, \cdots, x_n\}$:

$$\hat{\vartheta} := \arg\max_{\vartheta} Q_{x_1^n}(\vartheta) \tag{1}$$

The quality function used in [4] is:

$$Q_{x_1^n}(\vartheta) = Q_{x_1^n}(r, \theta, d) = \sum_{i=1}^{n} \max(q_{(r,\theta)}(x_i), \alpha q_{(r-d,\theta)}(x_i)) \tag{2}$$

where

$$q_{(r,\theta)}(x) = \max\left(0, 1 - \frac{d^2_{(r,\theta)}(x)}{\epsilon^2}\right) \tag{3}$$

The first term in the summation of Equation 2 calculates the contribution of a reference point x_i to the baseline, whereas the second term calculates the contribution of a reference point x_i to the descender line. Since a point can either lie on the baseline or the descender line, maximum of the two contributions is taken in the summation. Typically the value of α is set to 0.75, and its role is to compensate for the inequality of priors for baseline and descender such that a reference point has more chances to match with the baseline as compared to the descender line. The contribution of a reference point to a line is measured using Equation 3 and its value lies in the interval $[0, 1]$. The contribution $q_{(r,\theta)}(x)$ is zero for all reference points for which $d_{(r,\theta)}(x) \geq \epsilon$. These points are considered as outliers and hence do not belong to the line with parameters (r, θ). In practice, $\epsilon = 5$ proves to be a good choice for documents scanned at usual resolutions ranging from 150 to 600dpi. The contribution $q_{(r,\theta)}(x) = 1$ if $d_{(r,\theta)}(x) = 0$ which means the contribution of a point to a line is one if and only if the point lies exactly on the line.

The RAST algorithm is used to extract the text-line with maximum quality as given by Equation 1. Then all reference points that contributed with a non-zero quality to the extracted text-line are removed from the list of reference points and the algorithm is run again. In this way, the algorithm returns text-lines in decreasing order of quality until all text-lines have been extracted from the document image.

4   Evaluation and Results

To our best knowledge no dataset exists for questioned document examination evaluation. Therefore we generated some test images. Three images were generated by two different means: two images were generated by pasting a piece of paper containing a modified line of text over an existing one and then copying the document again using a multi function printer. The third image was generated by printing supplementary text on the original document. The three pages were extracted from the German version of the *Treaty establishing a Constitution for Europe*[1]. The original document images and the forged ones can be found in Figure 2 and Figure 3.

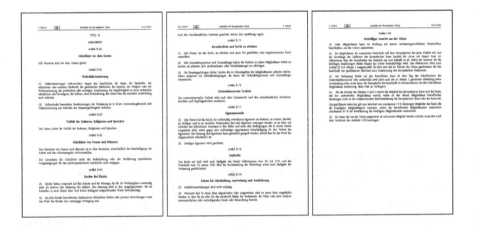

Fig. 2. Original pages

---

[1] http://eur-lex.europa.eu/JOHtml.do?uri=OJ%3AC%3A2004%3A310%3ASOM%3ADE%3AHT ML

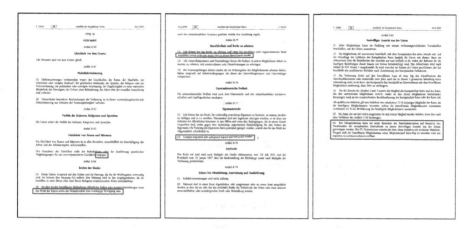

Fig. 3. Forged pages. On the first two pages lines have been altered by pasting new text over an existing line. In the third image a new paragraph was printed on an original document. The regions that have been modified are surrounded by a red box.

The training set consisted of 5 other document images from the treaty. All images were scanned with a resolution of 300dpi. For each line it was determined in what confidence interval its rotation angle lies. Lines in the 68% confidence interval remain white. The lines lying outside the 68% but inside the 95% interval are colored light red. The remaining lines are colored darker red for being still inside the 99.7% interval and totally dark red for being outside. Results for the three images be found in Table 1. For this Table, a line is considered as a fake line if it is outside of the 99.7% interval.

Table 1. Results of our approach for the three images. True positives are the lines that are faked and that have been detected as such. True negative are lines that have not been forged and that have not been reported as forged. False positive and false negative are the respective erroneous decisions.

Image	true positive	true negative	false positive	false negative
Image 1	1	26	2	1
Image 2	3	31	2	1
Image 3	0	23	0	5

A visualization of the results for the three images can be found in Figure 4, Figure 5 and Figure 6 respectively. It can be seen that false positives occurs quite frequently on short lines. On a short line the angle can be determined only quite roughly due to the discretization error. It can also be seen that printing text on an existing document may lead to alignments that are neither detectable by this method nor by manual inspection.

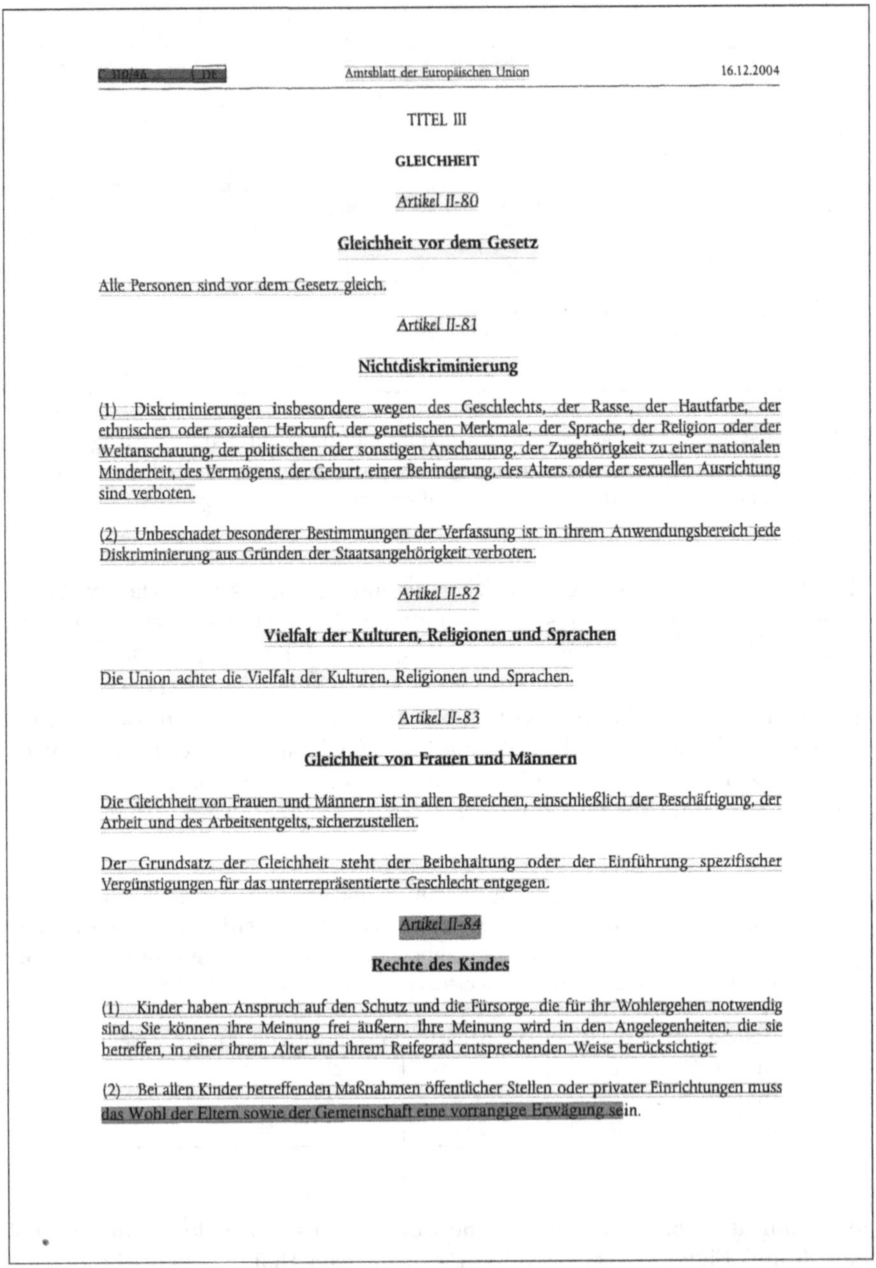

C 310/4A   DE   Amtsblatt der Europäischen Union   16.12.2004

TITEL III

GLEICHHEIT

Artikel II-80

Gleichheit vor dem Gesetz

Alle Personen sind vor dem Gesetz gleich.

Artikel II-81

Nichtdiskriminierung

(1)   Diskriminierungen insbesondere wegen des Geschlechts, der Rasse, der Hautfarbe, der ethnischen oder sozialen Herkunft, der genetischen Merkmale, der Sprache, der Religion oder der Weltanschauung, der politischen oder sonstigen Anschauung, der Zugehörigkeit zu einer nationalen Minderheit, des Vermögens, der Geburt, einer Behinderung, des Alters oder der sexuellen Ausrichtung sind verboten.

(2)   Unbeschadet besonderer Bestimmungen der Verfassung ist in ihrem Anwendungsbereich jede Diskriminierung aus Gründen der Staatsangehörigkeit verboten.

Artikel II-82

Vielfalt der Kulturen, Religionen und Sprachen

Die Union achtet die Vielfalt der Kulturen, Religionen und Sprachen.

Artikel II-83

Gleichheit von Frauen und Männern

Die Gleichheit von Frauen und Männern ist in allen Bereichen, einschließlich der Beschäftigung, der Arbeit und des Arbeitsentgelts, sicherzustellen.

Der Grundsatz der Gleichheit steht der Beibehaltung oder der Einführung spezifischer Vergünstigungen für das unterrepräsentierte Geschlecht entgegen.

Artikel II-84

Rechte des Kindes

(1)   Kinder haben Anspruch auf den Schutz und die Fürsorge, die für ihr Wohlergehen notwendig sind. Sie können ihre Meinung frei äußern. Ihre Meinung wird in den Angelegenheiten, die sie betreffen, in einer ihrem Alter und ihrem Reifegrad entsprechenden Weise berücksichtigt.

(2)   Bei allen Kinder betreffenden Maßnahmen öffentlicher Stellen oder privater Einrichtungen muss das Wohl der Eltern sowie der Gemeinschaft eine vorrangige Erwägung sein.

Fig. 4. Result for the first image. It can be seen that the single word forgery cannot be detected by this method, but the modified line at the bottom is recognized. Short lines tend to be problematic as the accuracy is just too low to estimate the rotation angle accurately.

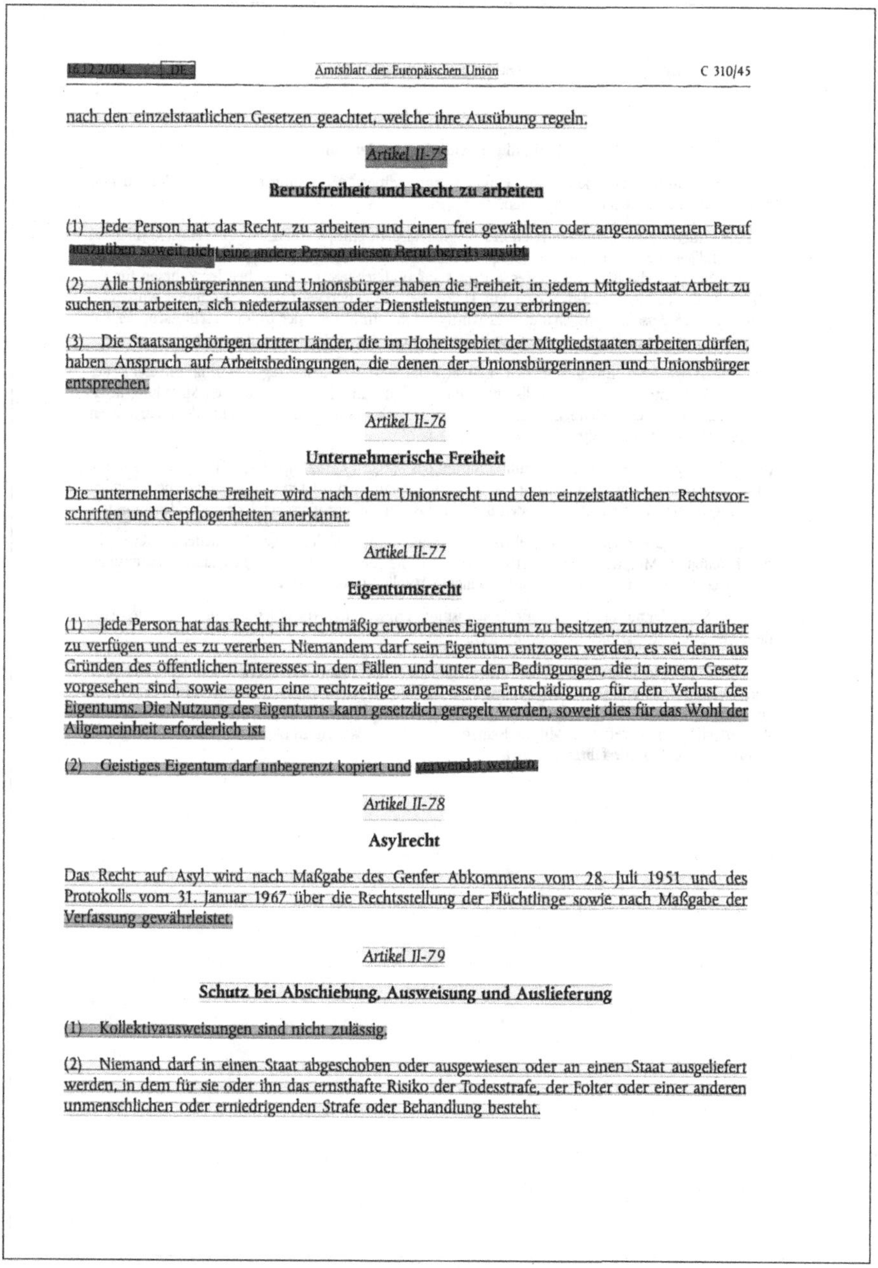

16.12.2004   DE   Amtsblatt der Europäischen Union   C 310/45

nach den einzelstaatlichen Gesetzen geachtet, welche ihre Ausübung regeln.

Artikel II-75

Berufsfreiheit und Recht zu arbeiten

(1) Jede Person hat das Recht, zu arbeiten und einen frei gewählten oder angenommenen Beruf auszuüben, soweit nicht eine andere Person diesen Beruf bereits ausübt.

(2) Alle Unionsbürgerinnen und Unionsbürger haben die Freiheit, in jedem Mitgliedstaat Arbeit zu suchen, zu arbeiten, sich niederzulassen oder Dienstleistungen zu erbringen.

(3) Die Staatsangehörigen dritter Länder, die im Hoheitsgebiet der Mitgliedstaaten arbeiten dürfen, haben Anspruch auf Arbeitsbedingungen, die denen der Unionsbürgerinnen und Unionsbürger entsprechen.

Artikel II-76

Unternehmerische Freiheit

Die unternehmerische Freiheit wird nach dem Unionsrecht und den einzelstaatlichen Rechtsvorschriften und Gepflogenheiten anerkannt.

Artikel II-77

Eigentumsrecht

(1) Jede Person hat das Recht, ihr rechtmäßig erworbenes Eigentum zu besitzen, zu nutzen, darüber zu verfügen und es zu vererben. Niemandem darf sein Eigentum entzogen werden, es sei denn aus Gründen des öffentlichen Interesses in den Fällen und unter den Bedingungen, die in einem Gesetz vorgesehen sind, sowie gegen eine rechtzeitige angemessene Entschädigung für den Verlust des Eigentums. Die Nutzung des Eigentums kann gesetzlich geregelt werden, soweit dies für das Wohl der Allgemeinheit erforderlich ist.

(2) Geistiges Eigentum darf unbegrenzt kopiert und verwendet werden.

Artikel II-78

Asylrecht

Das Recht auf Asyl wird nach Maßgabe des Genfer Abkommens vom 28. Juli 1951 und des Protokolls vom 31. Januar 1967 über die Rechtsstellung der Flüchtlinge sowie nach Maßgabe der Verfassung gewährleistet.

Artikel II-79

Schutz bei Abschiebung, Ausweisung und Auslieferung

(1) Kollektivausweisungen sind nicht zulässig.

(2) Niemand darf in einen Staat abgeschoben oder ausgewiesen oder an einen Staat ausgeliefert werden, in dem für sie oder ihn das ernsthafte Risiko der Todesstrafe, der Folter oder einer anderen unmenschlichen oder erniedrigenden Strafe oder Behandlung besteht.

Fig. 5. Result for the second image. The forged lines are curved in a way that the line find splits it into two lines and returns them as forged ones.

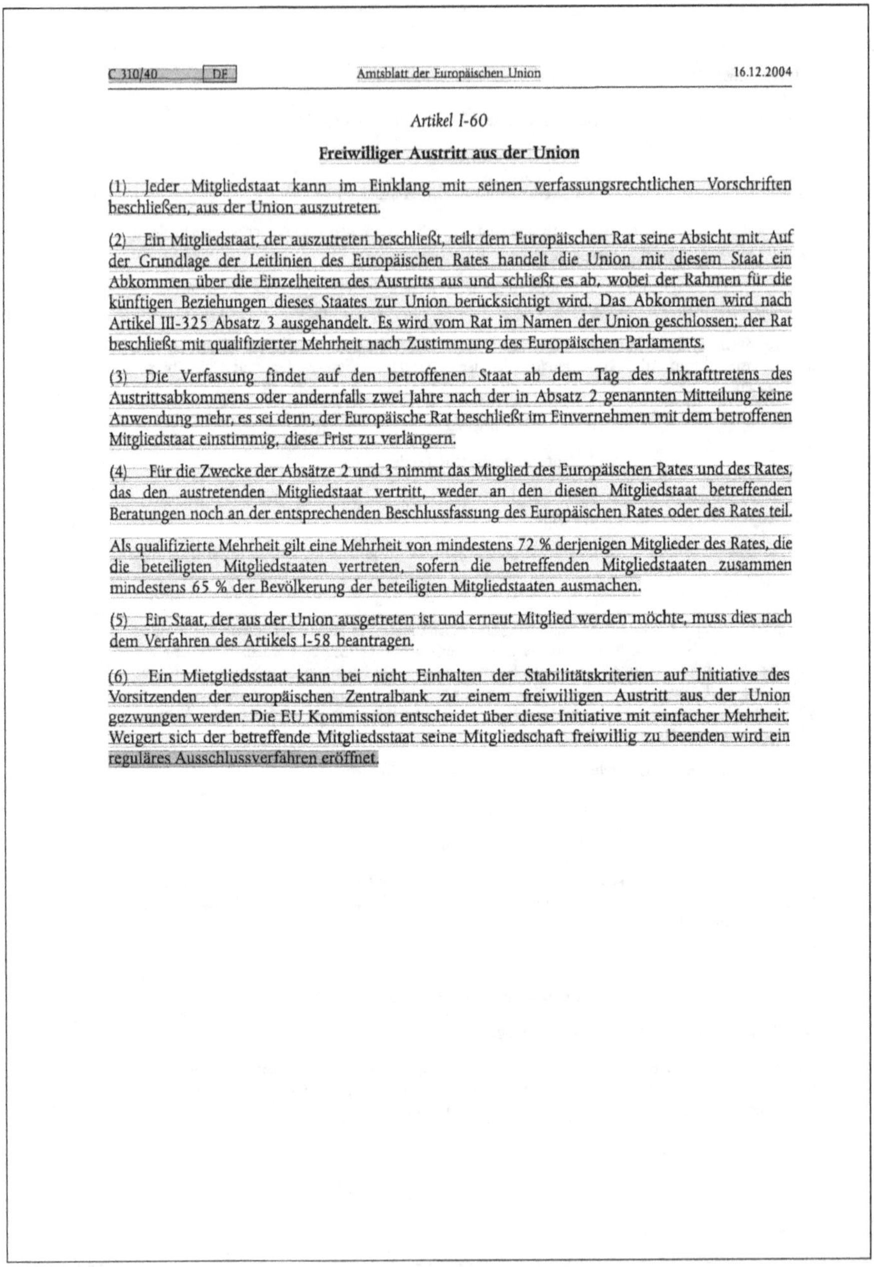

C 310/40　　DE　　　Amtsblatt der Europäischen Union　　　16.12.2004

Artikel I-60

Freiwilliger Austritt aus der Union

(1) Jeder Mitgliedstaat kann im Einklang mit seinen verfassungsrechtlichen Vorschriften beschließen, aus der Union auszutreten.

(2) Ein Mitgliedstaat, der auszutreten beschließt, teilt dem Europäischen Rat seine Absicht mit. Auf der Grundlage der Leitlinien des Europäischen Rates handelt die Union mit diesem Staat ein Abkommen über die Einzelheiten des Austritts aus und schließt es ab, wobei der Rahmen für die künftigen Beziehungen dieses Staates zur Union berücksichtigt wird. Das Abkommen wird nach Artikel III-325 Absatz 3 ausgehandelt. Es wird vom Rat im Namen der Union geschlossen; der Rat beschließt mit qualifizierter Mehrheit nach Zustimmung des Europäischen Parlaments.

(3) Die Verfassung findet auf den betroffenen Staat ab dem Tag des Inkrafttretens des Austrittsabkommens oder andernfalls zwei Jahre nach der in Absatz 2 genannten Mitteilung keine Anwendung mehr, es sei denn, der Europäische Rat beschließt im Einvernehmen mit dem betroffenen Mitgliedstaat einstimmig, diese Frist zu verlängern.

(4) Für die Zwecke der Absätze 2 und 3 nimmt das Mitglied des Europäischen Rates und des Rates, das den austretenden Mitgliedstaat vertritt, weder an den diesen Mitgliedstaat betreffenden Beratungen noch an der entsprechenden Beschlussfassung des Europäischen Rates oder des Rates teil.

Als qualifizierte Mehrheit gilt eine Mehrheit von mindestens 72 % derjenigen Mitglieder des Rates, die die beteiligten Mitgliedstaaten vertreten, sofern die beteiligten Mitgliedstaaten zusammen mindestens 65 % der Bevölkerung der beteiligten Mitgliedstaaten ausmachen.

(5) Ein Staat, der aus der Union ausgetreten ist und erneut Mitglied werden möchte, muss dies nach dem Verfahren des Artikels I-58 beantragen.

(6) Ein Mietgliedsstaat kann bei nicht Einhalten der Stabilitätskriterien auf Initiative des Vorsitzenden der europäischen Zentralbank zu einem freiwilligen Austritt aus der Union gezwungen werden. Die EU Kommission entscheidet über diese Initiative mit einfacher Mehrheit. Weigert sich der betreffende Mitgliedsstaat seine Mitgliedschaft freiwillig zu beenden wird ein reguläres Ausschlussverfahren eröffnet.

Fig. 6. Result for the third image. Inspection of the text-line rotation angles of the last paragraph have shown that there is no significant difference between the angles of this paragraph and the angles of the other paragraphs.

5   Conclusion and Future Work

In this paper we presented a method for helping the examiner to extract the line alignment feature that can be used to identify supplementary added content in a document. Using an automatic line finder and a statistical model for the distribution of the line angles, we are able to detect lines whose rotation angles differ from those of the other lines. Preliminary results show that the proposed method is able to identify these lines.

As the first results are promising but not yet carried on to a state of practical use, further work will have to be done. One important question is to analyze whether scanning in higher resolutions can improve the results or not. Furthermore, parameter estimation that is robust against outliers could be used to adapt the method to run on a single page and by this avoiding training of the method. Another possibility would be an interactive system where the examiner can choose the level of variation in line orientations that is allowed. A further interesting problem is the extension of the method to non-Latin script. Therefore the line model would have to be adapted to work on different scripts.

References

1. van Beusekom, J., Shafait, F., Breuel, T.M.: Document signature using intrinsic features for counterfeit detection. In: Srihari, S.N., Franke, K. (eds.) IWCF 2008. LNCS, vol. 5158, pp. 47–57. Springer, Heidelberg (2008)
2. Lindblom, B.S., Gervais, R.: Scientific Examination of Questioned Documents, pp. 238–241. Taylor and Francis, Abington (2006)
3. Shafait, F., Keysers, D., Breuel, T.M.: Performance evaluation and benchmarking of six page segmentation algorithms. IEEE Trans. on Pattern Analysis and Machine Intelligence 30(6), 941–954 (2008)
4. Breuel, T.M.: Robust least square baseline finding using a branch and bound algorithm, San Jose, CA, USA, January 2002, pp. 20–27 (2002)
5. van Beusekom, J., Shafait, F., Breuel, T.M.: Resolution independent skew and orientation detection for document images, San Jose, CA, USA, January 2009, vol. 7247 (2009)
6. Breuel, T.M.: A practical, globally optimal algorithm for geometric matching under uncertainty. Electronic Notes in Theoretical Computer Science 46, 1–15 (2001)
7. Breuel, T.M.: Implementation techniques for geometric branch-and-bound matching methods. Computer Vision and Image Understanding 90(3), 258–294 (2003)

Color Deconvolution and Support Vector Machines

Charles E.H. Berger[1,*] and Cor J. Veenman[1,2]

[1] Netherlands Forensic Institute
The Hague, The Netherlands
c.berger@nfi.minjus.nl
[2] Intelligent Systems Lab
University of Amsterdam
Amsterdam, The Netherlands
c.j.veenman@uva.nl

Abstract. Methods for machine learning (support vector machines) and image processing (color deconvolution) are combined in this paper for the purpose of separating colors in images of documents. After determining the background color, samples from the image that are representative of the colors to be separated are mapped to a feature space. Given the clusters of samples of either color the support vector machine (SVM) method is used to find an optimal separating line between the clusters in feature space. Deconvolution image processing parameters are determined from the separating line. A number of examples of applications in forensic casework are presented.

Keywords: forensic document analysis, questioned documents, color separation, color deconvolution, image processing, support vector machines.

1 Introduction

Machine learning methods offer two main promises for the forensic scientist. They can increase both the quality and the objectivity of their analysis. In this paper we use support vector machines [1,2] to determine the parameters for subsequent image processing to separate colors. Separating colors is a task often needed to improve the clarity of images when colors interfere with a feature of interest in the foreground or background. This can greatly enhance the visibility of faded, erased, or obscured features, or demonstrate that two written entries have a different color (suspected additions). For an overview of optical methods like IR/visible/UV luminescence and reflectance, or destructive methods such as thin-layer chromatography, high-performance liquid chromatography, and capillary electrophoresis see, e.g., Ref. 3.

2 Methods

Images were obtained by scanning the original documents with a high quality scanner (CreoScitex Eversmart Jazz). All the image processing and computational work was carried out in MATLAB® (The Mathworks, Inc., Bioinformatics Toolbox™).

---

* Corresponding author.

Z.J.M.H. Geradts, K.Y. Franke, and C.J. Veenman (Eds.): IWCF 2009, LNCS 5718, pp. 174–180, 2009.
© Springer-Verlag Berlin Heidelberg 2009

2.1 Color Deconvolution

In an earlier paper [4], color deconvolution was introduced as a way to achieve color separation, but an abbreviated explanation will be given here. In the RGB (red, green, blue) color-space we can see additive color mixing as the vector addition of RGB components to black. Equivalently, subtractive color mixing can be seen as the vector addition of CMY (cyan, magenta, yellow) components to white. For our purposes it is helpful to model the colors in an image as the vector addition of a desired (D) and un-desired (U) component to a background color ($\vec{p}$). The transformation of the RGB components to those components allows us to separate the desired and undesired components (with a vector $\vec{n}$ perpendicular to $\vec{u}$ and $\vec{d}$ so they span the 3D space):

$$\vec{c} = r \cdot \vec{r} + g \cdot \vec{g} + b \cdot \vec{b} = u \cdot \vec{u} + d \cdot \vec{d} + n \cdot \vec{n} + \vec{p} \text{ with } \vec{n} = \vec{u} \times \vec{d}. \tag{1}$$

By setting u to zero we remove the undesired component, after which we can transform back to RGB color space to find the new color $\vec{c}'$.

$$\vec{c}' = r' \cdot \vec{r} + g' \cdot \vec{g} + b' \cdot \vec{b} \equiv d \cdot \vec{d} + n \cdot \vec{n} + \vec{p} \tag{2}$$

The complex solution for $\vec{c}'$ is given in Ref. 4.

A similar calculation can be carried out when the purpose is not to remove a color, but to demonstrate a color difference. In this case the undesired and desired components can be shown in a false-color image to demonstrate the color differences and to evaluate how they correlate with features in the image.

2.2 Defining the Feature Vector

The analysis of the colors is based on the three-dimensional color histogram (see Figure 1). This histogram shows the distribution of all colors present in an image in the RGB color space. Histogram bins are represented by spherical halos that extend in proportion to the number of pixels with colors within that bin.

Note how the colors of the inks form elongated shapes, extending linearly from the large spherical cloud of colors associated with the paper background. This is due to differences in ink coverage in the pixels in and along the edge of the ink line. For our purposes these colors should share the same feature vector. Therefore, the spatial angles of the elongated cloud – with the center of the spherical cloud associated with the background as the origin – are a good choice for the feature vector. The feature vector of any color $\vec{c}$ in the image is given by the spatial angles x and y of the vector from the background color $\vec{p}$ to $\vec{c}$:

$$x = \text{atan2}(v_g, v_r) \quad \text{and} \quad y = \pi/2 - \text{atan2}(v_b, \rho),$$

$$\text{with } \vec{v} = \vec{c} - \vec{p} = \begin{bmatrix} v_r \\ v_g \\ v_b \end{bmatrix} \quad \text{and} \quad \rho = \sqrt{v_r^2 + v_g^2 + v_b^2}. \tag{3}$$

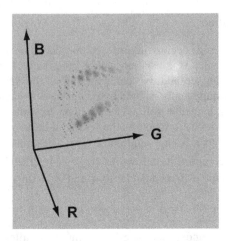

Fig. 1. 3D color histogram showing the colors of a blue and a black ballpoint ink on paper. Histogram bins are represented by spherical halos that extend in proportion to the number of pixels with colors within that bin.

2.3 Sampling Colors from the Image

It is possible to determine the background color and the undesired and desired components by sampling colors in 3 corresponding spots in the image [4]. In Ref. 5 these values were determined from the R, G, and B histograms of the complete image. For this paper we have sampled areas of the image by making masks manually, simply by painting over the image in an image editor. Instead of deriving our parameters for color deconvolution directly from the average color of sampled areas, we look at the pixels of the masked areas in the previously defined feature space. For clarity of presentation and to speed up the support vector machine calculations, the number of samples was limited to 200 per cluster (desired, undesired) by random selection. Using more samples did not change the results.

2.4 Support Vector Machine

In Fig. 3 an example is shown of an image of handwritten entries with different blue ballpoints on the left and right, with overlapping parts in the middle. Parts of the image with either ink and with the background paper were masked. After determining the average background color, all colors can be mapped into feature space. Fig. 2 shows 200 samples representative of either ink in feature space.

The support vector machine (implemented in MATLAB®) finds a linear maximum margin classifier: a straight line dividing both clusters with the items in the clusters as far away from the separating line as possible for separable clusters. For non-separable clusters the SVM method also finds the optimal separating line, minimizing the number of miss-classified items and their distance to the separating line.

The parameters for color deconvolution are given by 2 points in feature space, with the SVM separating line going perpendicularly through the middle of the line that connects them (see Fig. 2). For the present work the averages of both clusters were

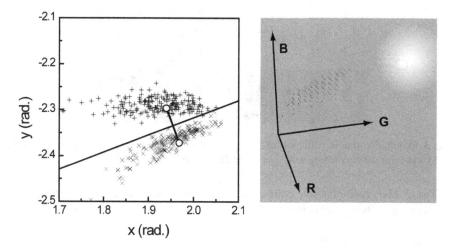

Fig. 2. The desired (+) and undesired (×) clusters associated with 2 blue ballpoint pen colors in an image (see Fig. 3). The SVM separation line is shown together with the feature vectors that form the parameters for the subsequent color deconvolution processing. On the right side the 3D color histogram shows both blue ink colors.

calculated, as well as the crossing point of the line connecting the cluster averages and the SVM separating line. From that crossing point the 2 points that give the color deconvolution parameters were determined by moving perpendicularly away on either side from the separating line over a distance equal to the average distance of both cluster averages to the separating line (see Fig. 2).

3 Results and Discussion

We will now apply the methods described above to a variety of images that require processing to either remove or discriminate color components.

3.1 Discriminating 2 Blue Ballpoint Inks

The first example concerns the differentiation of 2 very similar blue ballpoint inks, from a fraud case where an addition with a different pen was suspected. The SVM was trained with the left and right portions of the image (see Fig. 3), which gave the results earlier displayed in Figure 2.

For every example we'll show the separating line found by the SVM on top of a 2D histogram of the masked portions of the image (the scales of the 2D histogram are different every time; there are 100×100 histogram bins). The desired cluster's 2D histogram is added to the neutral gray background (going towards white) and the undesired part is subtracted (going towards black).

Similarly, after color deconvolution the separated desired and undesired components of the image are shown as lighter and darker than the neutral gray background respectively.

Fig. 3. Original writing in ballpoint ink, and both colors of blue ballpoint ink separated by the SVM and color deconvolution, as shown in the 2D histogram on the right

The 2 blue ballpoint ink colors are successfully separated and the processing clearly reveals the original entry as "C".

3.2 Address on a Label

In the following examples components will be differentiated with the purpose of re-moving the undesired component while maintaining as much as possible of the de-sired component. Figure 4 shows a portion of an address label from a fraud case that had been rendered impossible to read. Fortunately, the blue that covers the address is not exactly the same as that of the original blue handwriting.

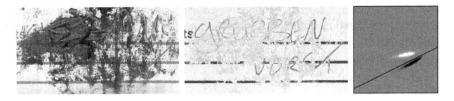

Fig. 4. A portion of an address label before and after processing

After processing the image shows the blue component associated with the original writing, except for some small regions where the covering blue was completely opaque. The name of the town of "GRUBBEN VORST" can now be read.

3.3 Handwriting on an Old Postcard

The paper of this old postcard had been damaged and discolored over the years. The parts masked were: some of the lightest areas as background; the dark discolorations as undesired; and the most visible fragments of the handwriting as desired.

The removal of the undesired component greatly enhances the visibility of the handwriting. The structural damage (visible as the light patches in the original im-age) is also much less apparent, though this improvement is of a more cosmetic nature since parts that were physically detached from the document can not be recovered.

Fig. 5. The handwriting on this old postcard was hard to read because of extensive yellowing in some parts and structural damage as well

3.4 Date Stamp in a Passport

The date stamp in this passport shows signs of mechanical erasure, making it very hard to read the date. The original image in Figure 6 is from a scan with amplified contrast, which amplified the stamp impression as well as the smudged background. Color deconvolution was used to selectively remove the smudge component of the image, while preserving the color of the stamp ink. The revealed text reads: "13 MARS 2006".

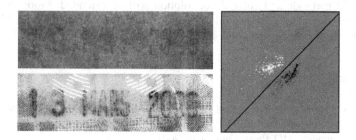

Fig. 6. Partly erased date stamp from a passport

While simply increasing the contrast does not enhance clarity by itself, it is a good first step because it makes sure the data in the image has a larger dynamic range.

3.5 Date Stamp on an Envelope

A stamp impression indicating the date of reception on an envelope was crossed out with a ballpoint pen, making it difficult to discern the date. Color deconvolution was used in combination with the SVM again to separate the color components of the stamp ink and the blue ballpoint ink, revealing the text: "INGEKOMEN 08 JULI 2004".

For this example the ballpoint lines from the original image are practically white in the processed image. Though preferably the background color should remain in those areas, it does enhance the clarity of the stamp impression.

Fig. 7. Crossed out stamp impression on an envelope

4 Conclusion

We introduced a method for estimating optimal parameters for color deconvolution using a linear support vector machine. With several examples from forensic casework, we showed that the method indeed gave good color separation results. Interfering colors were successfully removed, which made it possible to discern features of interest to the case.

In a next paper, the work on discriminating blue ballpoint inks with the combination of SVM methods and color deconvolution will be extended. More specifically, the value of the evidence for two inks coming from the same source or not [5] will be correlated with the success of color separation in the image processing.

References

1. Vapnik, V.: The Nature of Statistical Learning Theory. Springer, New York (1995)
2. Kecman, V.: Learning and soft computing: support vector machines, neural networks, and fuzzy logic models. MIT Press, Cambridge (2001)
3. Chen, H.S., Meng, H.H., Cheng, K.C.: A survey of methods used for the identification and characterization of inks. Forensic Science Journal 1, 1–14 (2002)
4. Berger, C.E.H., de Koeijer, J.A., Glas, W., Madhuizen, H.T.: Color Separation in Forensic Image Processing. Journal of Forensic Sciences 51, 100–102 (2006)
5. Berger, C.E.H.: Inference of identity of source using univariate and bivariate methods. Science and Justice (2009) (in Press) DOI:10.1016/j.scijus.2009.03.003

Author Index